PERGAMON INTERNATIONAL LIBRARY
of Science, Technology, Engineering and Social Studies

*The 1000-volume original paperback library in aid of education,
industrial training and the enjoyment of leisure*

Publisher: Robert Maxwell, M.C.

MONITORING TOXIC GASES IN THE ATMOSPHERE FOR HYGIENE AND POLLUTION CONTROL

Other Related Pergamon Titles of Interest

BOOTH: Industrial Gases

COOPER *et al*:   Particles in the Air (A Guide to Chemical and Physical Characterization)

* HUSAR: Sulfur in the Atmosphere

MEETHAM *et al*:   Atmospheric Pollution (Its History, Origins and Prevention) 4th Edition

STRAUSS: Industrial Gas Cleaning, 2nd Edition

TOMANY: Air Pollution: The Emissions and Ambient Air Quality

* *Not available under the Pergamon textbook inspection copy service.*

MONITORING TOXIC GASES IN THE ATMOSPHERE FOR HYGIENE AND POLLUTION CONTROL

By

WILLIAM THAIN

M.A., B.Sc., M.Inst.P.

formerly Research Co-ordinator,
The British Petroleum Company Ltd.

PERGAMON PRESS

OXFORD · NEW YORK · TORONTO · SYDNEY · PARIS · FRANKFURT

U.K.	Pergamon Press Ltd., Headington Hill Hall, Oxford OX3 0BW, England
U.S.A.	Pergamon Press Inc., Maxwell House, Fairview Park, Elmsford, New York 10523, U.S.A.
CANADA	Pergamon of Canada, Suite 104, 150 Consumers Road, Willowdale, Ontario M2J 1P9, Canada
AUSTRALIA	Pergamon Press (Aust.) Pty. Ltd., P.O. Box 544, Potts Point, N.S.W. 2011, Australia
FRANCE	Pergamon Press SARL, 24 rue des Ecoles, 75240 Paris, Cedex 05, France
FEDERAL REPUBLIC OF GERMANY	Pergamon Press GmbH, 6242 Kronberg-Taunus, Pferdstrasse 1, Federal Republic of Germany

First edition 1980

British Library Cataloguing in Publication Data

Thain, W
Monitoring toxic gases in the atmosphere for hygiene and pollution control. - (Pergamon international library).
1. Air - Pollution
2. Gases, Asphyxiating and poisonous
3. Environmental monitoring
I. Title
614.7'12   TD885   79-41300
ISBN 0-08-023810-6

In order to make this volume available as economically and as rapidly as possible the author's typescript has been reproduced in its original form. This method has its typographical limitations but it is hoped that they in no way distract the reader.

Printed in Great Britain by A. Wheaton & Co. Ltd., Exeter

*To my wife, without
whose care and encouragement this book
would not have been written*

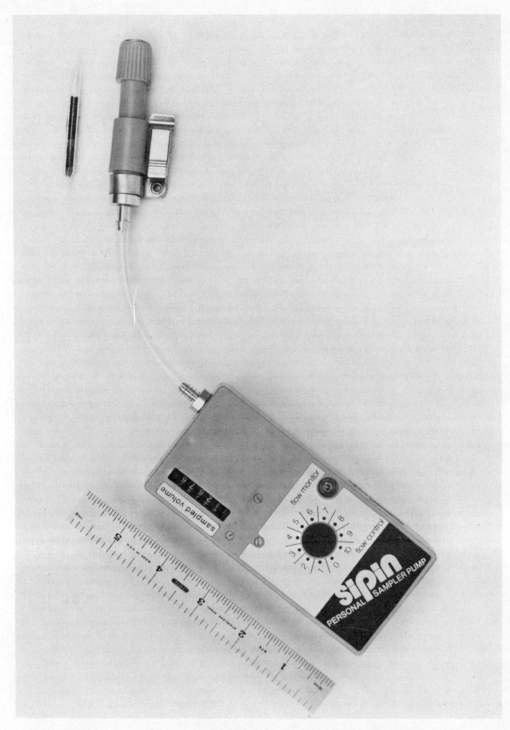

Frontispiece. Personal Sampling Pump with charcoal absorption tube.
(Courtesy of Sipin International)

CONTENTS

LIST OF ILLUSTRATIONS

PREFACE

The awareness of the hazards presented by exposure to toxic gases and vapours has increased, and analysts are now frequently required to make measurements to monitor the presence of such substances in the atmosphere. The term "monitoring" has two meanings in the United Kingdom. It has a wide sense in which it means the repeated measurement of concentration of pollutants so that changes can be followed over a period of time. In a more restricted sense it means the determination of the value of the concentration of a pollutant in relation to some standard in order to judge the effectiveness of a system of control.

In both senses the monitoring of pollutants requires sound and reliable methods of sampling and measurement of gases and vapours in the atmosphere, both within the working area of industrial plants and in the environment outside them. Standard methods for making measurements of toxic substances in the atmosphere can be obtained in the publications of national regulatory authorities, for example Health and Safety Executive in the United Kingdom and National Institute for Occupational Safety and Hygiene in U.S.A. These methods, however, are analytical procedures which are reliable if followed strictly but may not always be appropriate for a particular requirement.

This book presents the principles of atmospheric monitoring and indicates where difficulties can occur, and where errors in sampling and measurement may be expected. It should be useful to analysts who are meeting this type of problem for the first time, and also to analysts who have to alter established methods or to develop new methods to meet their own particular problems and circumstances. It does not detail methods for measuring or sampling specific substances except in so far as these illustrate a particular problem. Similarly where instruments or devices are described it is not implied that these are unique or are the best available. They have been selected from those readily available in the United Kingdom, most European countries and North America, and probably obtainable in most other countries where atmospheric monitoring is undertaken. They have been chosen to illustrate points of design, operation or procedure which are important in atmospheric monitoring.

It has been assumed throughout that the reader is familiar with the conventional techniques of analysis found in present-day laboratories and so details of the analytical procedures have been omitted except where special details are important to the analysis of samples from atmospheric monitoring programmes.

In general, units of, or acceptable for use in conjunction with, the International System (SI) have been used in the text. An exception to the use of such units occurs where reference is made to important engineering standards of equipment which are quoted by manufacturers in terms of Imperial units.

ACKNOWLEDGEMENTS

Although my first attempts at atmospheric monitoring took place over thirty years ago the contents of this book are largely drawn from experience and information gained during more recent times when my responsibilities as Research Co-ordinator with the British Petroleum Company included the development of analytical techniques for monitoring purposes.   During part of this time I was also Convenor of a specialist group set up by the Chemical Industries Association to advise on and develop monitoring methods for vinyl chloride.   Many R & D and control laboratories throughout the BP Group took part in both of these programmes, and laboratories in other U.K. PVC companies contributed extensively to the second.   I acknowledge freely the work, both development and application, done by many laboratory and works staff, which laid the foundations for this book.

I wish to thank the manufacturers of monitoring equipment many of whom have provided me with detailed information about their equipment and its application for use in the book.   Amongst those who have been particularly helpful are, in alphabetical order, the following companies or their U.K. agents: C.F. Casella, Century Systems, Draeger Safety, Environmental Monitoring Systems, Foxborough Analytical Ltd., MDA Scientific, D.A. Pitman Ltd., Sipin International, Universal Environmental Instruments Division of J. and S. Sieger, and 3M United Kingdom Ltd.

I also express my thanks to the many other laboratories, both industrial and governmental, in U.K. and overseas, with whom I have exchanged information and experiences on monitoring problems.   In particular I wish to mention the NIOSH Laboratory in Cincinnati, U.S.A. who provided me with many reports of their work and permitted me to quote from them in this book.

Finally, I wish to thank Miss M.K. Tolley who with great patience produced the first typescript of the text from a very rough manuscript, and Dr. D.J. Moore who read the draft and made a number of suggestions which have been incorporated to the benefit of the final version.

W. Thain.

Chapter 1

TOXIC HAZARDS AND THEIR MEASUREMENT

SPECIFICATION OF HAZARDS

It is now accepted that every industrial user of chemical substances has a duty to protect his workers and neighbours from any adverse effects of the substances used in his processes. Volatile substances may enter the atmosphere from leaks in equipment, spillages during handling, plant malfunctions and by way of effluent gases and liquids. These atmospheric pollutants may give rise to hazards both to those in the working areas and to local residents. The user must therefore be aware of the toxic hazards associated with these substances and must also be aware of the amount of each which is present in the atmosphere as a result of his operations. The methods of measuring the concentrations of these toxic substances and of comparing the measured with specified permitted values is the subject of this book.

It is obvious that certain chemicals create hazards to health because exposure to sufficiently high concentration of these substances for even very short periods causes adverse effects and, in some cases, may be fatal. Such acute toxicity is easily recognised and precautions against exposure to these substances are usually readily implemented. It has become apparent, however, that there are substances which do not give rise to adverse effects after short exposure but which cause damage to the health of persons exposed to them for long times. This hazard is much more difficult to recognise because of the delay in the appearance of the adverse effects and the difficulty of correlating these effects with exposure to the substance. If the effect is to cause an unusual disease, such as the angiosarcoma of the liver caused by prolonged exposure to vinyl chloride, the correlation is more readily observed. Otherwise statistical methods for demonstrating the correlation must be adopted. There are differences in individual sensitivities to chemical substances and age, sex, physical condition, and physical demands imposed by work, all affect the extent to which any person is endangered by any specific substance. Further, the hazard presented by a mixture of toxic substances is often different from that created by each substance individually and synergism between substances is possible.

Lists of dangerous materials and poisons and their lethal doses are available. These, however, are limits for short exposures and so additional information is required on the concentrations of gases and vapours, not necessarily acutely toxic, to which exposure can be tolerated for a long time without adverse effects on health.

The first attempt to prepare systematic information on the tolerance of humans to long term exposure to toxic substances was made in 1933 when the American Conference of Governmental Industrial Hygienists published a list of maximum workplace concentrations for a number of substances.   The Committee based their findings on the best data available from toxicology and industrial medicine and it reviewed and supplemented the list each subsequent year as more data became available.   By 1947 maximum workplace concentrations had been established for 150 gases and vapours and the list was published in tabular form.   Subsequently in 1950 the Conference introduced the concept of "Maximum Allowable Concentration" to replace the name "Maximum Workplace Concentration" previously used.

The Maximum Allowable Concentration is the maximum concentration of a substance which can be tolerated without adverse effects to the health of a human being.   The tolerance limit was arrived at only after the possibility of noxious effects had been excluded following industrial observations, and after examinations in industrial applications covering a number of years or, alternatively, following extensive animal experiments.   The exposure period is based on an 8 hour day.   The Maximum Allowable Concentration values for gases and vapours were calculated in parts per million by volume (ppm) or in milligrams per cubic metre, based on a temperature of 20°C and barometric pressure of 760 torr.   The existence of a Maximum Allowable Concentration value in the list implies that an analytical method is available to make measurements down to that level of concentration.

It was later recognised that the Maximum Allowable Concentration values were the limits of tolerance of the various substances and not the concentrations that should be permitted in industrial applications.   In consequence, to avoid any implication which might result from this name, the Maximum Allowable Concentration values were renamed "Threshold Limit Values" (TLV) in 1962.   The list of TLVs is now published annually by the American Conference of Governmental Industrial Hygienists (Ref. 1). These values are in common use and many countries republish them automatically as national values.   In other countries, including the United Kingdom, they are reviewed before being adopted.   In view of their widespread use the significance and limitation of Threshold Limit Values are discussed later in this Chapter.

While these limits were being assigned to chemical substances in U.S.A., similar work was being undertaken in Germany and U.S.S.R. which resulted in two other national lists of concentrations.   In Germany the concept of Maximale Arbeitsplatz - Konzentration (MAK - Maximum Workplace Concentration) was adopted in 1954 and lists of the values are published (Ref. 2).   The MAK is defined as the concentration of an industrial agent in the workplace atmosphere which, in the light of current knowledge, does not normally damage the health of persons employed at that place even if they are exposed over a lengthy period, normally for 8 hours a day but not exceeding 45 hours a week.   The values are decided by a Committee, the Kommission zur Prüfung Gesundheitschadlichen Arbeitsstoffe (Committee for Testing of Noxious substances) mainly in the light of experience in industrial health and hygiene.   Less weight is given to the results of experimentation with animals.   Information from other organisations is considered.   In addition to the MAK values the Committee publishes a list of "Technische Richtkonzentration" (TRK, Technical Guideline Concentrations).   This is a guide to users of substances for which the medical evidence is insufficient to enable MAK values to be established.

In U.S.S.R. the concept of "Maximum Permissible Concentration" (MPC) within the industrial environment has been adopted and the values for over 500 substances were published in 1974 (Ref. 3).   The MPC is based on a 45 hour working week but the medical tests and considerations on which the figures are assigned are entirely different from those used in the preparation of the U.S.A. and German lists and so the values, in certain cases, differ greatly from those found in the TLV and MAK lists.   In consequence, when the World Health Organisation attempted in the early

1960s to prepare a list of maximum workplace concentrations which would be inter-
nationally acceptable, it was unable to do so because of the divergence between the
national values and the methods by which they had been arrived at.   However, in
1969, the concept of "Maximum Permissible Concentration" was agreed upon, based on
an exposure during an 8 hour day and 5 day week, and the values were published for
24 substances used in the chemical industry.   The unit of concentration adopted
was milligrams per cubic metre.

THRESHOLD LIMIT VALUES

Purpose and Limitations

Threshold Limit Values are the limiting values of concentration used for the control
of exposure to toxic substances in many countries including the United Kingdom, and
so the definition, purpose and limitations of these values must be fully understood.

The Threshold Limit Value (TLV) is the time-weighted average concentration to which
a normal person may be exposed, day after day for 7 to 8 hours over a 40 hour week,
without adverse effect.   The values are based on the most reliable information
currently available and incorporate a considerable safety margin.   The time-weighted
average permits excursions above the TLV provided they are compensated by equivalent
excursions below the limit, but the extent and duration of excursions permitted are
related to various factors which depend on the nature of the substance concerned.
For certain substances a ceiling value has been designated and is indicated by the
letter C in the TLV list.   This value is an absolute limit and should not be exceeded even for a short time.

The basis on which values have been established differs from one substance to
another.   In some cases the main criterion may be toxicity but in others irritation
or narcosis may be the determining factor.   For this reason TLVs do not necessarily
indicate the relative toxicity of different substances.

No inference should be drawn from the omission of a substance from the TLV list.
It may be inocuous, but equally it may be a substance for which insufficient data
are available for a TLV to be assigned.   Alternatively, it may be a substance to
which special considerations apply, for example, it may have been shown to be
carcinogenic.   Acceptable concentrations of exposure to carcinogenic substances are
difficult to determine because little is known of the dose-effect relationship.

The word "skin" is added to values for certain substances in the TLV list.   This
does not indicate that the substance has any adverse effect on the skin itself, but
that the substance can be absorbed through intact skin, particularly by direct contact.   Such absorption is additional to that from inhalation of the vapour and so
the airborne concentration does not, by itself, give a sufficient indication of the
danger arising from the substance.

The evaluation of health hazards from mixtures of substances is a complex subject.
Unless information is available to the contrary, or expert advice is available, it
should be assumed that the effects of the various components are additive.

Although it is believed that regular exposure to a concentration of the Threshold
Limit Value of a substance will not result in injury to health, it is good practice
to reduce the exposure of workers to much lower values wherever practicable.

Units of Concentration

The concentrations of substances dealt with in toxicity control are usually very low and are expressed either as parts per million by volume (ppm) or as weight per unit volume, usually mg/m^3, on a basis of a temperature of 20°C and barometric pressure of 760 torr. Concentrations can be converted from one method of expression to the other provided the gram molecular volume of the substance is known.

When lower levels of concentration are being quoted the expression "parts per billion" (ppb) has been used. The use of this unit is not recommended because the definition of the unit "billion" differs from one country to another. Usually the American definition of billion, i.e. one thousand million, is implied but the unit, if used should always be defined.

Application of Threshold Limit Values to the General Environment

Threshold Limit Values refer to concentrations of harmful substances in the work-place and the values have been determined in such a manner as to relate to conditions of exposure there. They are, therefore, not appropriate for evaluating hazards to people who live around factories. However, in the absence of other more appropriate standards, some guidelines based on TLVs have been suggested for evaluating these risks. It has been proposed that concentrations of toxic substances in inhabited areas, sometimes called "immission concentrations", should not exceed one thirtieth of the TLV. This factor is arrived at by assuming that inhabitants of such areas are exposed continuously, i.e. about three times the exposure time used in defining TLVs, and with a further factor of one tenth to make allowance for the fact that residents include the aged, infirm and infants, all of whom may be expected to be more susceptible to the hazards.

In Germany the concept of Maximale Immission Konzentration (MIK - Maximum Immission Concentration) has been introduced. This is the maximum concentration of substances which can be present in the atmosphere without resulting in injury to health or discomfort to any persons, including infants and the sick, even though they are exposed to it continuously. The MIK values for a number of substances are listed by the Verein Deutscher Ingeniere (Association of German Engineers). It has been proposed that an MIK value of one twentieth of the MAK value should be used as a provisional figure for organic compounds which have not been listed.

The calculations of acceptable limits by the use of TLVs or MAKs and the suggested factors give values which are useful as general guides. Much lower values should always be aimed at, especially when dealing with substances which are known to present special hazards.

THE ANALYTICAL PROBLEM

Information Required by Management

Management may require various types of information on concentrations of toxic substances to ensure that workers and local residents are not subjected to hazards from the chemicals involved in the factory processes. This is in addition to information required for other aspects of safety, e.g. fire protection, and for process control purposes. At different times they may require, for their own use or for provision to authorities, the following types of information, each of which presents different problems to the analyst.

General survey. Measurements may be required to determine what toxic materials are present in the atmosphere and the approximate concentration of each. These measurements are required to provide a description of the state of the atmospheric environment, either inside or outside the factory, and its changes.

For this purpose the analytical method must provide for the collection of a sample
which truly represents the atmosphere and from which no components are lost.   This
has to be followed by a procedure for the identification of the various components
and measurement of their concentrations.

Specific substance monitoring.   The concentration of a specific toxic substance in
the atmosphere in the presence of other chemical substances may be required.

This measurement may be made by a technique which measures a property specific to
the substance whose concentration has to be determined.   Unfortunately, few techniques are completely specific and some allowance or adjustment for interferences
due to other substances have to be applied.   Alternatively, the determination may
be made by a non-specific technique of measurement, provided that prior to the
analysis the required substance is separated from others which would interfere with
the measurement.

It is common for such measurements of concentrations to be required to determine if
regulations have been complied with.   The method of measurement may be specified
in the regulations in which case the conclusions can, strictly speaking, only be
drawn by measurements made by that procedure.   However, analysts use other methods
more convenient to them provided that the method adopted gives the same result as
the specified method.   The different methods may, of course be subject to different
errors and interferences and so the analyst must be completely aware of the characteristics of both methods before he can use a substitute method with confidence.

Unspecified substance monitoring.   In monitoring for explosion hazards it is not
necessary to determine the concentration of individual substances but only whether
a hazard of explosion exists.   This can be done by measurement of the physical
property of flammability.   There is no method of measuring general toxicity and so
a measurement analogous to flammability cannot be made.   However, there are
occasions when it can be assumed that only one atmospheric pollutant is present.
This may be because only one substance is present in a process, or because the
problem is such that the identity of the measured substances is irrelevant to the
action to be taken as a result of the measurement.   Such a situation arises in
leak detection and certain measurements required as a result of spillages.   In
such cases rapidity of measurement is usually an important requirement and simple
measurement techniques of a non-specific nature may be used.   However, the analyst
must be aware of potential interferences and errors if he is to interpret the results
correctly.

Accuracy, Sensitivity and Limit of Detection

The analytical methods chosen must be capable of providing data with the accuracy
required and must be able to detect the concentrations which exist in each application.   To show compliance with a standard the method should have high sensitivity
at the concentration specified in the standard (sensitivity is the slope of the
calibration curve) in order that the result should be unambiguous.   The statistical
criteria for showing non-compliance with a standard are dealt with in Chapter 9.
The limits of detection of the methods must be appropriate to the problem being
undertaken.   Concentrations in the atmosphere around the factory are much lower
than those in the working area and so methods used for measurement in the external
environment must have a lower limit of detection than is necessary for methods used
to monitor the working area.   This may be achieved by using methods of higher

sensitivity but the limit of detection of a method of measurement may be set by factors other than its sensitivity, for example by reagent blanks or by interference from other substances.   Thus, the limit of detection of a method may vary from one type of sample to another, in particular, from laboratory samples prepared for calibration and method evaluation to samples taken in the field.   For this reason great care should be taken in reporting the results of measurement at levels of concentration near the limit of detection.   If a substance is not detected by an analytical technique it should be reported as "Not detected" followed by the limit of detection of the method for that type of sample.   If the limit of detection is omitted the information is clearly incomplete.   The alternative form of report "Less than xxxppm", where xxx is the limit of detection of the method, implies confirmation of the presence of the substance and should not be used unless such confirmation has been obtained.

For certain purposes confirmation of the identity of the substance measured may be required.   As the limit of detection is approached positive identification becomes increasingly difficult and may only be possible at a level of concentration one or two orders of magnitude greater than the limit of detection.

Duration of Measurement

Two types of measurement of concentration are required in most studies on atmospheric pollution.   One is aimed at measuring the concentration of a specified substance in the atmosphere at a particular instant, and the other is required to measure the time-weighted average concentration over a specified time.   This time may vary from 5 or 15 minutes for measurement of short excursions or for demonstrating compliance with ceiling values, to the 8 hours specified by most working area standards, or to even longer times specified for special purposes.

Instantaneous concentrations.   If the concentration of the substance is varying with time it is very difficult to make a true measurement of its concentration at any instant.   Most measuring devices require a finite amount of sample to be collected on which to make the measurement, though in certain cases this sample may be very small.   If the concentration varies over the time the sample is being collected, the concentration in the sample will differ from the instantaneous concentration at the time of measurement by an amount which depends on the rate and extent of the change.   The sampling time depends on the type of measurement technique in use and may be as short as a few seconds for certain direct reading instruments up to several minutes for slow acting measuring devices.   This problem is considered in more detail in Chapter 3.

Time-weighted average concentrations.   Time-weighted average concentrations (TWA) over a period of time can be measured directly by two methods.   In the first method a monitor is used to measure continuously the concentration of the substance, and the time-weighted average is calculated from the record of concentrations produced. If the monitor is such that discrete samples have to be taken for analysis the measurement is not truly continuous but, provided the samples are taken frequently and are separated by only short intervals of time, the time-weighted average measured in this way will not be significantly in error.

In the second procedure the time-weighted average is obtained by collecting a sample of the atmosphere at a constant rate during the time over which the average is required.   The sample is then analysed and the resulting concentration is the required average.   The limiting factor in the accuracy of such measurements is usually the constancy of the rate of flow of the sample over the time required.

Techniques and equipment suitable for this type of measurement are described in Chapters 5, 6 and 7.

The continuous measurement and continuous sampling methods both involve fairly elaborate equipment which has to work continuously in the area where measurement is required.   Such equipment is not always available and then an alternative procedure based on measurement of instantaneous concentrations at random times may be used. In this method the distribution of values of concentration with time has to be determined or assumed.   Then, provided sufficient samples are taken at truly random times, the probable value of the time-weighted average can be calculated statistically.   If no information on the distribution of concentration values with time is available it is usual to assume a log-normal distribution.   The procedure for measuring the time-weighted average in this way is described in Chapter 9.

Characteristics of Specific Types of Monitoring

The locations and duration of monitoring depend on such factors as the toxicity of the substances involved, the types of operations being carried out, the variations in concentrations which occur in normal working and the probability of process malfunction giving rise to concentrations higher than normal.

Working area monitoring.   If highly toxic substances are being handled it may be desirable to monitor continuously within the working area the concentrations at or near the stations where workers are located.   This area monitoring can be carried out by fixed monitors located within the working area, or by a monitor attached to a sampling system which collects samples from points within the working area.   The characteristics of such sample collection systems are considered in Chapter 3.

In handling less toxic substances or very low concentrations of more toxic substances, and where stable working conditions exist giving rise to concentrations well within a safe region and with few excursions, it may be sufficient to monitor the general working area by occasional random sampling methods using either direct reading instruments or grab sampling techniques.

Personal monitoring.   The area monitoring procedure provides an average measurement of the concentration throughout the working area but does not provide data on the average concentration to which any individual worker is exposed.   This can be done only by equipping the worker with a personal monitoring device which he can carry while he is working.   Most of the systems of personal monitoring currently in use provide the measurement at the end of the working shift and do not, therefore, provide any warning of high concentrations as they occur.   For this reason area monitors are frequently used in conjunction with personal monitors.

Systems of personal monitoring have been reviewed (Ref. 4) and consist mainly of the techniques mentioned above for measurement of time-weighted average concentrations.

Area survey and leak detection.   Portable devices, preferably direct reading instruments should be available for monitoring concentrations which arise from accidents such as spillage, and for checking before entry, the concentrations in un-monitored areas.   In these cases rapid measurement is highly desirable but accuracy of measurement is usually of secondary importance.   If direct reading devices, either instrumental or colour indicating, are not available it is necessary to resort to sample collection methods.

A special case of this type of measurement is leak detection. Leaks are found by observing the concentration gradient of the leaking substance in the atmosphere and by following the increase in gradient to the source of the leak. Thus the device used must be able to detect the substance concerned over a wide range of concentrations to provide the response required at the beginning of the operation without being overloaded in the final stages. It must be rapid and precise in operation but absolute accuracy of measurement and specificity to the substance concerned are of less importance.

The environment around the factory. The concentrations of toxic substances in the atmosphere in the neighbourhood of the factory can be measured by procedures similar to those used to monitor the working areas. However, the concentrations are likely to be very much lower than those within the factory and so sampling and measurement techniques must be chosen accordingly, and because measurements have to be made at points remote from the factory, self-contained equipment with its own power supply, capable of operating unattended, may be required.

The usual requirement is to determine the extent to which effluents from the factory are adding to the burden of toxic substances in the atmosphere. Where the factory concerned is the only one in the neighbourhood in which a particular substance is being used, it can reasonably be assumed that the total amount of that substance in the atmosphere results from the operations of that factory. However, if other factories nearby are also emitting the same substance it is difficult to determine to what extent each factory contributes without measuring the actual emissions. This problem is dealt with in the following section.

The concentrations of a substance in the atmosphere around a factory vary from place to place and from time to time depending on weather conditions, wind speed and direction, local topography, plant operations, etc. The substance may be emitted from a single point in the factory or from a number of scattered sources which may not be constant in emission. The interpretation of the results of measurements of these atmospheric concentrations is complicated and has been the subject of mathematical studies which are beyond the scope of this book (for example Ref. 5). However, certain simple empirical laws on the distribution with time of concentrations of atmospheric pollutants are referred to later in this chapter.

Factory effluents and emissions. Where a substance is emitted from a single source, e.g. a fume stack, the total quantity emitted may be determined by measuring the flow rate of the stack gases and the concentration of the pollutant. The measurement of concentration, however, can present problems because of the difficulty of obtaining a representative sample of gases flowing in a duct. ˙A sample taken from a point near the centre of a straight section of duct, remote from bends or junctions, may for most purposes be assumed to be representative of the contents of the duct, provided the gas flow is turbulent. If the flow is laminar multiple sampling techniques must be employed. (Ref. 6.).

The sampling probe must be designed in such a way as to provide the measuring or sample collection device with a flow of sample at the correct flow rate and pressure, taking account of pressure effects caused by the flow of gas into or across the end of the probe. A sample can be collected over a time to determine the total amount of the specified substance which passes through the duct in that time. If the total flow in the duct varies during that time of sample collection the sample should be taken, not at a constant rate, but at a rate which is varied to be maintained proportional to the total flow rate in the duct. Where the emissions occur from a number of adjacent points in a processing plant they can be gathered by a temporary or permanent extraction hood and ducted away for measurement

as described above.   If, however, the emissions occur from a number of points
which are widely separated, and which are not easily monitored individually, it is
necessary to calculate the emission from its effect on the external atmospheric
environment.   The method, called "up-wind, down-wind measurement", involves determination of the pollutant burden added to the atmosphere by the factory by measuring
the difference in pollutant concentration of the air approaching the factory (up-
wind) and that leaving the factory (down-wind).   The measurements must be made
when the wind is constant in velocity and in a direction in which the pollutant
cloud is least likely to be disturbed by topographical features.   The speed of the
wind affects the size of the pollution cloud and the distribution of material within
it.   Measurement must be made when the wind speed is within the limits which can
be allowed for in the data reduction calculations.   The method does not affect
factory operations and sampling is continued for a period long enough to smooth out
cyclic variations of emission due to process operations.

One measurement point is usually adequate in the up-wind direction but five or more
may be required in the down-wind direction depending on the assumptions made on the
concentration pattern within the pollutant cloud.   It is usually assumed that the
concentrations vary according to a Gaussian law in both horizontal and vertical
directions.   The actual pattern can be determined by making measurements at selected points, but measurements in a vertical direction may be difficult to make.   The
results of the various measurements can then be used in diffusion equations to
calculate the total emission from the factory.

The procedures for determining the optimum location of measurement points and for
calculating the results are described in a report by the Environmental Protection
Agency of U.S.A. (Ref. 7).

Liquid effluents from processes may contain volatile toxic substances in solution
which may leave the effluent and enter the atmosphere.   There are a number of
established methods for measuring low concentrations of volatile organic substances
in aqueous solution (e.g. Refs. 8, 9) but these are not within the scope of this
book.

Recording of Data from Fixed Analysers

Fixed analysers may be used to make measurements, either continuously or in rapid
succession, of the concentrations of toxic substances in the atmosphere of plants
processing these substances.   Similarly they may be used at remote sites to measure
concentrations of pollutants in the atmospheric environment.   In both cases instruments fitted with automatic sampling devices can produce large quantities of data
which are difficult to handle by manual methods.   Fortunately, most instruments
used in such applications provide an output signal in electrical form.   The
simplest method of displaying the information is on a strip chart recorder, either
as a continuous trace or in bar chart form.   Such a display indicates readily the
trends in the changing value of concentration, and the recorder system can be made
to operate an alarm if the value moves outside a predetermined range.   Such a
system is satisfactory for the display of a single parameter, but when it is used
to display several parameters, e.g. concentration of several components in a sample,
or the concentration of a single pollutant in several samples analysed in succession,
the chart tends to become confused with the overlapping data and trends become
difficult to observe.   Again, when actual values are required for any reason, e.g.
calculation of time-weighted averages, or recording of peak values, the chart has
to be measured at each point where measurement is required, and correction and
calibration factors have to be applied before the calculation can be made.   This
is a laborious and time consuming operation.   Furthermore, when data are being
collected from remote instruments there is a limit to the distance over which

analogue output signals may be transmitted to operate a recorder.

The electrical output signals from monitoring instruments can readily be collected by a data processing system based on a small electronic computer.   Within the system is an analogue-to-digital converter in which the signals are changed to digital form, and this digital signal can then be processed in the standard methods of computing.   The signal can be corrected for known errors and converted to a concentration value using a calibration relation contained within the system.   This value can be stored for future use, can be displayed either automatically or on demand, can be used in the calculation of averages as required, can be sorted into bands for distribution analysis, and can be used to trigger signals or alarms. (Refs. 10, 11).

Where measuring points are remote from the data processing equipment each point is usually equipped with an analogue-to-digital converter.   The resulting digital signals are easily transmitted over long distances, without deterioration in accuracy, over lines or radio links, using standard signal transmission techniques.

Data processing systems can be used to generate control signals which may be required to operate valves in sampling lines and in gas chromatograph analysers.   They can command samples of known concentration to be inserted to check the operation of the analyser, and can adjust the calibration and correction parameters according to the measurements made on these standard samples.   When samples from different points are being measured it is usual for the samples to be introduced into the analyser for measurement in a predetermined sequence.   However, a system has been described in which the samples from the various points are measured in a sequence which depends on the past history of the concentrations at the various points.   (Ref. 12). Thus, if one sample point shows a trend of increasing concentration and is approaching a critical value, measurements are made more frequently at that point than at others where concentrations are well below the critical value.

Safety of Electrical Equipment

Many of the techniques of measurement and sampling used for monitoring toxic gases in the atmosphere involve the use of electrically operated equipment.   These measurements may have to be made in areas which are subject to special regulations because of the existence of flammable gases.   Then the equipment used must be specially designed and constructed to avoid the possibility of causing ignition of these gases, and must comply with the regulations in force in these areas.

Unfortunately, the regulations which cover the design requirements for equipment for use in hazardous areas differ from one country to another, although they are now being harmonised within the European Economic Community.   However, it cannot be assumed at present that equipment which is in use in one country in this type of application is acceptable elsewhere.

Sampling and Measurement Procedures

Throughout this chapter it has been assumed that a sample can be obtained which is truly representative of the atmosphere which is to be monitored and that there is a means of measuring the concentration of the component specified.   The sampling and measuring systems must be matched to give the sensitivity, accuracy, specificity, speed of response and sample pattern required.

Chapter 2 illustrates how certain methods of measurement have been used in environmental problems, making use of their characteristics to meet the requirements of the problem. Chapter 3 describes the difficulties which can be encountered in presenting samples to these measuring devices or to sample collectors. Chapter 4 describes some of the factors which are important in the application of the technique of measurement of low concentrations by colour change devices, which has long been popular for this purpose. Chapters 5, 6 and 7 describe the various techniques and equipment which are available for collecting samples, over short or long times, for laboratory analysis. Chapter 8 illustrates methods of calibrating and testing monitoring procedures and Chapter 9 outlines certain statistical aspects of atmospheric monitoring. Chapter 10 considers some possibilities for the future development of monitoring systems.

REFERENCES

1. Threshold Limit Values, American Conference of Governmental Industrial Hygienists, Cincinnati, Ohio, U.S.A.

2. Kommission zur Prüfung Gesundheitsschadlicher Arbeitstoffe, Uberprufung von MAK - Werten, Arbeitsschutz Nr. 12, 381 (1974).

3. I.V. Sanotsky, D.I. Mendelev Journal of the All-Union Chemical Society, 19, 125 (1974).

4. W. Thain, Measurement of Exposure to Toxic Gases by Personal Monitors, Report on International Symposium on the Prevention of Occupational Risks in the Chemical Industry Frankfurt-am-Main 21-23 June 1976. Berufsgenossenschaft der Chemischen Industrie, Heidelberg, Germany.

5. E.V. Somers, Dispersion of pollutants emitted into the atmosphere, Chapter 1 of Air Pollution Control, Part I, (1971). Edit. Werner Strauss, Interscience, New York.

6. Flow and Gas Sampling Manual, Report No. EPA 600/2 - 76 - 203, July 1976, Environmental Protection Agency, Research Triangle Park, North Carolina, U.S.A.

7. Technical Manual for Measurement of Fugitive Emissions: Upwind/Downwind Sampling Method for Industrial Emissions, Report EPA 600/2 - 76 - 089a, April 1976, Environmental Protection Agency, Research Triangle Park, North Carolina, U.S.A.

8. B. Kolb, Application of an automated headspace procedure for trace analysis by gas chromatography, J. Chromatography, 122, 553 (1976).

9. T.A. Bellar, J.J. Lichtenberg, & J.W. Eichelberger, The determination of vinyl chloride at microgram/litre level in water by gas chromatography, Environmental Sci. & Technol. 10, 926 (1976).

10. G.L. Baker & R.E. Reiter, Automatic systems for monitoring vinyl chloride, Amer. Ind. Hyg. Assoc. J. 38, 24 (1977).

11. M.A. Field & R.C. Moore, A Computer Controlled Multi-point Sampling and Measuring System for Monitoring Gaseous Contaminants in the Atmosphere. International Conference on the Monitoring of Hazardous Gases in the Working Environment, London, 12 - 14 December 1977.

12. Resin Plants Meet OSHA's Vinyl Chloride Targets, Oil and Gas J. 24, May 10, 83, (1976).

Chapter 2

TECHNIQUES OF MEASUREMENT

CHOICE OF TECHNIQUE OF MEASUREMENT

Many methods of measurement have been used in atmospheric monitoring both in direct reading devices and in laboratory instruments used to analyse samples which have previously been collected.   In common with most analytical techniques they consist of the measurement of some physical property of the substance which is being monitored, sometimes after it has been subjected to some prior treatment to ensure that it is in the correct chemical or physical state for the measurement to be made. The choice of measurement technique for any particular application depends on such factors as the type of sample presented, whether or not the equipment is required to be portable, the concentration range to be covered, the sensitivity of measurement required, and whether substances which would interfere with a method of measurement are likely to be present.

Most problems fall into one of two groups, one in which a measurement has to be made directly on the atmosphere which is being monitored, and the other in which a sample of the atmosphere may be taken and analysed subsequently, usually in the laboratory. More direct methods using simpler equipment are usually employed in the first type of measurement, whereas the full range of analytical techniques found in the laboratory can be used for the second type.

In practice the final choice of method often depends on the availability of equipment and expertise.   Some of the specialised equipment which has been used in problems of atmospheric monitoring is costly and not found in all industrial laboratories.   If it is essential to use these techniques it may be necessary to employ the services of an analytical consultant laboratory which operates the specialist equipment concerned.   Further, the problems of measurement at very low levels of concentrations are considerable.   Samples must be handled with extreme care to avoid loss or contamination which would be insignificant in handling samples at higher levels of concentration, and assumptions made in normal analytical work on, for example, efficiency of extraction procedures and linearity of instrument calibration, may have to be questioned.   Such measurements, therefore, should be entrusted only to staff experienced in this type of work.

Examples of techniques of measurement which have been used in the determination of toxic gases in the atmosphere are given in the following section, together with some information on those characteristics of each technique which are important in monitoring applications.   The techniques are subdivided into groups according to the

12

type of property which is the basis of measurement.   No attempt has been made to
quote figures for the sensitivity or the limit of detection of the various techniques.   The former is very dependent on the details of the design of the actual
equipment and, in some cases, on the substance being measured.   The latter is
dependent on the sensitivity and stability of the equipment, and on external factors
such as interference from other substances which may be present.   The effect of
these factors on the limit of detection is discussed in Chapter 3.

A number of the techniques of measurement which are described can be used in different ways in toxic gas monitoring.   For example, many of the techniques can be
used in detectors for gas chromatography.   In this application they are required
to measure the relevant substance in the presence only of the carrier gas (assuming
that separation is efficient).   Some of the techniques, however, can be used to
measure the specified substance directly in the atmosphere without separation.   A
device which is based on such a technique and designed to operate in this way is
described in this Section as an "atmospheric monitor".

MEASUREMENT TECHNIQUES AVAILABLE

Techniques Based on Ionisation

Flame ionisation.   When an organic compound is burned in a hydrogen flame, ions
are produced in numbers which depend on the concentration of the compound in the
combustion gases.   These ions can be collected by an electrode which is maintained
at high potential, and the electric current which results is proportional to the
ion concentration.   This electric current is very small and special electronic
circuits and components are required to amplify it sufficiently to operate an
indicating instrument.   In particular, the electrical insulation of the electrode
must be very good otherwise currents due to electrical leakage will interfere with
the measurements.

A detector based on flame ionisation responds to all substances which produce ions
on combustion.   This includes practically all organic compounds, but it will not
detect, and is unaffected by, water vapour, nitrogen oxides, carbon monoxide and
carbon dioxide.   The signal produced is proportional to the ion concentration
over a very wide range which depends on the design of the detector, but which can
exceed five decades.   The response of the device to any particular substance depends on the number of ions formed in the combustion process.   This, in turn,
depends on the chemical nature of the substance and so it is necessary to calibrate
the detector using a sample containing a known concentration of the substance,
though with experience the response to a given substance can be estimated.   In
general, a flame ionisation detector is more sensitive to hydrocarbons than to
other types of organic compounds.   Compounds which contain oxygen show a lower
response than is observed for hydrocarbons, and this reduction is particularly
marked for compounds which have a high ratio of oxygen to carbon.   Compounds which
contain nitrogen behave in a manner similar to those containing oxygen.   Compounds
which contain halogens also show a lower response than that given by hydrocarbons,
and those which contain no hydrogen give the lowest response.

Because the detector is sensitive to all types of organic compounds its operation
can be affected by dust or combustion products which condense within the combustion
chamber.   These can give rise to ions, sometimes in an irregular manner, which
result in an ion current even when no sample is being flowed into the detector.
To avoid this, the combustion and sample gases should be filtered to remove dust,
and the combustion system should be kept clean.

The principle is widely used in detectors for gas chromatography but is also used

in atmospheric monitoring devices.   In this latter application the device detects
not only atmospheric contaminants, but also the methane which occurs naturally in
the atmosphere at concentrations up to about 5 ppm.   Correction must be made for
the signal due to this atmospheric methane, and this is one of the factors which
limits the lower level of detection of the device.

Photo-ionisation.   The technique of measurement by photo-ionisation is similar in
general principle to that of flame ionisation but the processes of ionisation
involved are different, and this leads to the techniques having different characteristics.   In photo-ionisation the ionisation is caused by irradiation with
ultra-violet light.   If the radiation has an energy greater than the ionisation
potential of the substance which is being irradiated, that substance will be
ionised and the ions can be collected and measured in a manner similar to that used
in the flame ionisation technique.   Many organic substances have ionisation potentials which are lower than those of the major components of the atmosphere and so,
by selecting an ultra-violet source which emits photons of an appropriate energy,
it is possible to ionise and detect these organic substances in the atmosphere.
Similarly, if certain components of a gas mixture have ionisation potentials which
exceed the energy of the ionising source, these will not be detected, irrespective
of their nature.   The device, therefore, shows some selectivity and in this respect
differs from flame ionisation.

While the ionisation potential serves as a rough guide as to whether or not a given
substance will show a response to the photo-ionisation device, it does not predict
the magnitude of the response.   The sensitivity of the technique for any given
substance depends on the chemical structure of the substance and the conditions
under which ionisation takes place.   For example, if organic substances are
irradiated in the presence of molecular oxygen by photons of energy greater than
the ionisation potential, the yield of ions is reduced by a quenching effect of
the oxygen, even though it is not itself ionised.   Fortunately the magnitude of
this quenching effect is independent of the concentration of oxygen at atmospheric
composition, and so slight changes of oxygen concentration do not affect the
sensitivity to a specified substance of a device based on photo-ionisation, if it
is calibrated with known concentrations of the substance in air.   However, the
sensitivity of the device to the same substance in an atmosphere of nitrogen would
be much greater.   Water vapour also shows a quenching effect.

The linearity of response depends on the design of the ionisation cell but can
cover a range exceeding three decades.

The technique can be used both in a detector for gas chromatography and in an
atmospheric monitor.

Thermionic detection.   In thermionic detectors ionisation is induced thermally,
either by a flame or by an electrically heated filament.   If a salt of an alkali
metal is located adjacent to the source of ionisation, the ion current which results
from ionisation of compounds which contain nitrogen, phosphorus or halogens is much
greater than would be observed in the absence of the alkali metal.   Several
mechanisms have been suggested to explain the phenomenon but none fully accounts
for all aspects of the observed behaviour of the devices based on the principle.

The selectivity and sensitivity of the devices are highly dependent on the geometry
of construction, temperature of ionisation, nature and purity of the alkali metal
salt, and the operating parameters.   The sensitivity is also dependent on the
chemical structure of the compound which is being measured.

The technique is used in selective leak detectors designed for use directly in the
atmosphere, and also in a selective type of detector for gas chromatography.   In
its latter form it is a modification of the flame ionisation detector, but does not
have the stability and reliability of the flame ionisation device.   The thermionic
detector requires regular maintenance and calibration and must be used under constant conditions.   It should be used, therefore, only by experienced analysts.

Mass spectrometry.   In the mass spectrometer the sample is bombarded under standard
conditions by a beam of electrons.   Under these circumstances any substance will
crack in a reproducible manner and give rise to a unique pattern of fragments of
different masses, which depends on the molecular structure of the substance.   The
pattern can be observed, and the abundance of the fragments of each mass can be
measured, by separating the fragments according to mass.   This is achieved by
accelerating the fragments electrically and separating them in a magnetic field
(magnetic sector instrument), or in a mass filter (quadrupole instrument) after
which the abundance of fragments of each mass is measured in turn by a photon multi-
plier.   The whole operation is carried out under conditions of high vacuum.

The mass spectrum of a mixture of substances is the sum of the spectra of the
individual components each at its own concentration.   It is possible, therefore,
to analyse mixtures by resolving the complex spectrum of the mixture into the
simpler spectra of its components.

Mass spectrometers are versatile, highly sensitive and highly selective instruments
and are normally used in the laboratory because of their complexity, though instruments specially made for application on industrial plants, usually in process
control systems, are available.

Electron capture.   This technique is allied to ionisation methods.   The electron
capture device consists of a cell in which are located two electrodes, one of which
contains a radioactive source of nickel-63 or tritium.   The radiation ionises the
air within the cell and an ionisation current can be produced by maintaining a
potential difference between the electrodes.   When an electronegative species is
introduced into the detector it captures electrons and reduces the standing current.
The extent of this decrease depends both on the number of electron-capturing species
present and on their electronegativity.   The detector, therefore, will measure the
quantity present.   It is selective for highly electronegative species, such as
compounds which contain halogens, oxygen and unsaturated groupings.   The device is
useful as a gas chromatograph detector of extremely high sensitivity for certain
halogenated compounds, but it has limited dynamic range of measurement.   It is
adversely affected by the presence of water vapour and so cannot be used as an
atmospheric monitor.

Techniques Based on Thermal Measurement

Heat of combustion.   This technique measures only combustible gases.   The sample
is passed over a catalytic sensor which is heated to a temperature above the
ignition temperature of the gas which is to be measured.   A filament within the
sensor is connected in an electrical bridge circuit.   The heat of combustion
changes the resistance of the filament and unbalances the bridge by an amount dependent on the concentration of the combustible gas in the sample.

The technique is generally non-specific although some selectivity can be provided
by selection of the temperature of operation of the filament.   The technique is
commonly used for monitoring explosion hazards but is rarely sufficiently sensitive

for use in detecting toxic hazards.

Thermal conductivity.   The sample is passed through a cell which contains a heated
wire filament.   The loss of heat from this filament, which depends on the thermal
conductivity of the various components of the sample, is measured electrically.

The technique is non-selective and is affected by normal atmospheric components
such as water vapour and carbon dioxide.   It is useful for detecting high concentrations of gases but has limited application as a measuring technique at the
concentration levels encountered in toxic gas monitoring.

Techniques Based on Chemical and Electrochemical Reactions

Chemiluminescence.   When organic substances undergo certain chemical reactions,
light is emitted with an intensity which cannot be attributed to thermal radiation.
The most common reaction which gives rise to this effect is the reaction of ozone
with unsaturated compounds.   The wavelength of the light which is emitted depends
on the nature of the substances involved, and the intensity of the light is dependent on their concentrations.   This forms the basis of a specific and highly sensitive detector commonly used for monitoring ozone and oxides of nitrogen.

Electrolytic conductivity.   This technique operates by treating the sample in such
a manner that the substance which is to be measured is converted to an ionic
species, which is dissolved in deionised water.   The change in electrical conductivity of the water is a measure of the concentration of the ionic species, from
which can be calculated the concentration of the original substance in the gas
sample.

Compounds which contain sulphur, halogens or nitrogen can be determined by this
technique.   Selectivity depends on the specificity of the chemical reactions which
produce the ionic species.   A supply of deionised water of constant quality is
required.   This is usually provided by circulating the water from the conductivity
cell through an ion exchange column.   Constant conditions of operation, temperature
and concentration of other ionisable substances must be maintained to ensure satisfactory performance.

This technique is used both in an atmospheric monitor and in a detector for gas
chromatography.

Coulometry.   The substance which is to be measured is made to react stoichiometric-
ally with a reagent which is generated electrolytically.   The current passed to
the coulometric generator to maintain equilibrium is a measure of the substance
undergoing reaction.   Compounds which contain sulphur, nitrogen or chlorine can be
determined in this manner after they have been converted to a suitably reactive
chemical form.   A potentiometric sensing circuit produces an error signal which,
after amplification, controls the generation of reagent to adjust it to the rate of
arrival of the analyte.

The selectivity of the technique is entirely dependent on the specificity of the
preliminary conversion and the coulometric reactions.   A device based on this
technique can be used for atmospheric monitoring or as a detector for gas chromatography.   It requires no calibration with mixtures of known concentration provided
stoichiometry is achieved.

<u>Electrolytic reaction</u>. The molecules of the gas which is to be measured are made to enter an electrolyte where they dissociate and take part in an electrochemical reaction, thereby generating an electric current. The magnitude of the current is proportional to the number of gas molecules which undergo reaction. A different type of electrolyte is required for each substance and so the device is specific for the substances which will undergo reaction in the electrolyte in use.

Techniques Based on Optical Measurement

<u>Colorimetry</u>. Many substances can be detected and measured at low concentrations by their reaction with reagents which produce coloured products. This technique is very widely used and the characteristics of systems based on colorimetry are described in detail in Chapter 4.

<u>Interferometry</u>. The refractive index of a mixture of gases depends on the refractive indices of its various components. A contaminant which has a refractive index different from that of air can be detected and measured in air by measuring the difference between the refractive index of the sample and that of uncontaminated air, by comparison of interference fringe patterns.

The sensitivity of the technique depends on the difference in refractive indices of the gases. The method is unspecific and is subject to interference from all other gases which differ in refractive index from those being measured. Carbon dioxide and water vapour interfere markedly with the measurement of hydrocarbons.

<u>Spectral absorption</u>. When radiation is passed through a sample which contains organic gases or vapours, absorption of radiation will occur at certain wavelengths. The type of substance present in the sample can be determined by the wavelengths at which absorption takes place, and the quantity of the substance can be calculated from the amount of radiation absorbed, using the Beer-Lambert law.

The measurement is usually made using monochromatic light and the wavelength used may lie in any part of the optical spectrum from the ultra-violet region suitable, for example, for detecting low concentrations of mercury vapour, to the infra-red. Absorption in the ultra-violet region is also useful for the detection of certain types of organic compounds, for example, aromatic hydrocarbons and unsaturated compounds, which can be detected at very low levels of concentration. In the infra-red region all organic compounds have unique spectra and can be determined by absorption measurements. However, many of the absorptions are weak and only the strongest absorption bands are useful for analytical purposes.

The monochromatic radiation required for analysis can be provided by a selective line source, for example a vapour discharge lamp, or can be isolated from a source of continuous radiation by means of optical filters, prisms or gratings. The selectivity of the measurement depends on the optical band-width of the beam of radiation. Filters tend to produce a wider range of wavelengths and so instruments using filter monochromators tend to suffer from interference from substances which absorb at wavelengths adjacent to that being used for analysis.

Because the absorptions are weak it is necessary to provide a long absorption path length in order to measure substances at low concentration.

The absorption spectrum of a sample can be measured by an interferometric technique instead of by measurements at single wavelengths isolated from a source of infra-red radiation. The interferogram is converted to a spectrum by Fourier transform calculations which can be done rapidly by an electronic computer connected on-line to the interferometer.   This technique, which requires costly equipment, improves markedly the speed of response, the resolution of absorption bands, and the sensitivity of the measurement.

The sensitivity of spectral measurements can also be increased by using the technique of laser optoacoustic spectroscopy.   A discretely tuneable gas laser is used as a source of radiation.   This radiation is modulated at an acoustic frequency and passed through a cell which contains the sample.   The radiation absorbed by the sample heats the gas on each cycle and a pressure transducer measures the periodic pressure variation within the cell.   The equipment is costly but the sensitivity of measurement is extremely high.

Tuneable lasers are also used as infra-red sources for devices which make measurements on atmospheric pollution remote from the analyser by measurement of back scattered radiation (LIDAR) and also in instruments which make measurements of infra-red absorption over very long path lengths of several kilometres.

Flame photometry.   Certain elements are readily detected by the light which they emit when their compounds are heated or undergo combustion.   For example, when sulphur compounds are burned in a hydrogen-rich flame, they emit light at a wavelength of 394 nm with an intensity approximately proportional to the square of the concentration of sulphur in the sample.   The light can be measured by a photomultiplier tube and an optical interference filter is usually included in the optical path to exclude light of other wavelengths.

This technique is used in a gas chromatography detector which detects only sulphur compounds and also in an atmospheric monitor.

Techniques Based on Effects on Semi-Conductors

Solid state sensors.   These devices are based on non-stoichiometric metallic oxides whose electrical properties depend on the atmosphere which surrounds them.   In simple forms they are sensitive but unspecific, and respond to almost all organic substances with which they come in contact.   However, by doping the semi-conductor it is possible to increase the selectivity to certain gases.   Early designs based on comparatively large detecting elements showed hysteresis in following concentration changes.   Later designs based on transistor technology use much smaller sensing elements and are rapid in response.

Techniques Based on Separation

Gas chromatography.   Many of the techniques of measurement described above are non-specific and so can be used only if correction can be made for the effects of other substances which may interfere with the measurement of the required substance. However, if this substance can be separated from the other substances which are present, the measurement can be made by a non-specific technique.   The most common technique used for the separation of individual components in atmospheric monitoring is gas chromatography.   This is used both in laboratory analysis of collected samples and as a portable monitor or field analyser.

Liquid chromatography.   For certain problems where separation of the components cannot readily be made by gas chromatography it may be necessary to use other techniques.   The most common of these is liquid chromatography but other techniques such as thin layer chromatography may have to be employed in some analyses.

EXAMPLES OF MEASUREMENT TECHNIQUES APPLIED IN MONITORING

This section describes the application of various measurement techniques in problems of atmospheric monitoring. The list is not exhaustive and the examples described have been chosen to show how the methods have been selected to use the characteristics of the technique employed on the solution of the particular problem.

Portable Monitors and Leak Detectors

The main requirements of a portable device for monitoring and leak detection are speed of measurement, portability, which demands robustness, low power consumption and minimum use of other supplies, for example, gases or compressed air. Specificity requirements depend on the extent to which substances other than that which is to be monitored are likely to be encountered, and the range of measurement required depends on the particular application. In general, leak detectors require less specificity but greater range than monitors.

Portable monitors. The rapid response and sensitivity of flame ionisation have led to its use in portable monitors. (Ref. 1). The instruments have been designed to operate from a battery and to include their own supply of hydrogen in a lightweight cylinder. The flame is protected by flame traps to ensure that it will not ignite flammable gases in the atmosphere.

The lack of selectivity of the technique is a limitation in certain cases and, if monitoring is required in areas where a number of organic gases are present in the atmosphere, a separation prior to measurement is required. Separation sufficient for many purposes can be achieved by incorporating a short separating column in the sampling tube. In this form the instrument becomes effectively a simple gas chromatograph, but, because the column does not operate under controlled conditions, the separation is not necessarily complete. Nevertheless, it can be sufficient to enable reliable measurements to be made of atmospheric pollutants in the presence of other organic substances. (Ref. 2).

Photo-ionisation has been made the basis of a portable monitor. (Ref. 3). Again interference from other substances may limit its application but, by correct choice of the energy of the ionising radiation, the technique can be given some selectivity. In particular, by selecting an energy of less than 12 eV the device is made insensitive to the methane which occurs naturally in the atmosphere, but it will still respond to many organic compounds. Thus the device can be used to measure very low concentrations of organic gases in the atmosphere, without allowance having to be made for the effect of methane as is required in a flame ionisation detector.

A portable infra-red absorption analyser which uses a multiple-pass cell giving an absorption path length of 16 m, has been developed as a portable monitor. The instrument contains its own rechargeable battery and requires no other services or supplies. The measurement is made at a single wavelength which is isolated by an optical filter, and which co-incides with a strong absorption band in the spectrum of the substance which it is required to monitor. (Ref. 4).

Semi-conductor sensors have been incorporated in portable monitoring devices. These monitors are usually unspecific but one manufacturer claims that his sensors are selective and can be made specific to any one of 150 compounds. Performance data to substantiate this claim is not available at the time of writing. (Ref. 5).

The most widely used portable detector and monitoring system is that based on colour change, either in the form of indicator tubes or sensitised paper tape.

Examples of these devices are described in Chapter 4.

Leak detectors.   The instrumental portable monitors described above can all be used as leak detectors but it is not practicable to use normal colour change systems for this purpose because their time of response is too great.   In addition, a number of other techniques can be used for finding leaks.   A thermal ionisation device which is highly sensitive to halogenated compounds but which is not calibrated accurately in terms of concentration of a specified substance is used for detecting leaks of halogenated gases, especially refrigerant gases. (Ref. 6).   A device which makes a measurement of thermal conductivity at high sensitivity is designed to compare the concentration of a gas at two adjacent points from which samples are taken simultaneously.   The measurement made is of concentration gradient of the required substance and this gradient can be traced back to the source of the leak. (Ref. 7).   The detector is used in this comparative manner to eliminate the effect on the measurement of the presence of atmospheric water vapour.   It is reasonably assumed that the water vapour concentration is the same at the two points from which samples are taken.   Because of the magnitude of the effect of water vapour it is not possible to make absolute measurements of concentration by thermal conductivity.

Fixed Monitors in the Working Area

Working area monitors require measurement techniques which have high sensitivity at the appropriate level of concentration, reasonable speed of response, and selectivity to the substance whose concentration is to be measured.   Unlike portable monitors, weight, size and consumption of power and other services are secondary considerations.

The techniques used for measurement in portable monitors can also be used in working area monitors.   An area monitoring system based on flame ionisation detection is available commercially (Ref. 8), although in most monitoring systems flame ionisation detectors are used in conjunction with gas chromatography columns to improve selectivity. (Ref. 9).   A photo-ionisation device designed as a fixed monitor is available. (Ref. 10).

Spectroscopic techniques have been widely used as detector systems in area monitoring.   A filter type infra-red spectrometer has been used for monitoring a wide range of toxic gases (Ref. 11), and a grating spectrometer has been specially designed for the same purpose. (Ref. 12).   Measurements have been made very rapidly with an area monitor in which the detector is a Fourier Transform Infra-Red Spectrometer. (Ref. 13).   High speed measurements have also been made with a system based on a quadrupole mass spectrometer (Ref. 14).   A flame photometer detector system is used in an analyser which measures total sulphur in the atmosphere. (Ref. 15).

Detection systems based on chemical and electrochemical techniques have been used in working area monitors.   Chemi-luminescence detectors are commonly used for the measurement of oxides of nitrogen in the atmosphere. (Ref. 16).   Monitoring systems for vinyl chloride have been based on combustion or oxidation of the vinyl chloride, absorption of the products of reaction in water, and measurement of the electrical conductivity of the water. (Ref. 17).   A similar system in which the combustion stage is replaced by a process of degradation by ultra-violet irradiation has been described. (Ref. 18).

Colour change detectors used in area monitoring applications are usually based on sensitised paper tape.   Such a system is used for monitoring the concentration of toluene di-isocyanate. (Ref. 19).

Personal Monitoring

Personal monitors are required to be sufficiently compact and light in weight to
enable them to be carried by the person whose exposure is being monitored, and
they must measure the average concentration over the time of exposure, often eight
hours.   This implies a technique which makes little demand on electrical power.
The sensitivity required and the specificity of measurement depend on the particular
substance being measured and the likelihood of the presence of other substances
which may interfere with the measurement.

A pocket size gas detector and alarm system for gases such as hydrogen sulphide,
phosgene and nitrogen dioxide has been developed. (Ref. 20).   The gas diffuses
through a membrane, which acts as a selective filter, into an electrolytic cell in
which a current is generated.   The current is displayed on a calibrated meter
and activates an alarm at a predetermined value.

A dosimeter for carbon monoxide is based on a fuel cell sensor.   A pump moves air
over the sensor and the value of the accumulated exposure is stored electronically.
To read this value the dosimeter is connected to a support console unit which
contains a reading and display circuit. (Ref. 21).

Direct measurement methods based on colour change reactions have also been used in
monitoring systems.   According to type, these provide integrated exposure values
or exposure profiles.   These systems are described in Chapter 4.

Because of the problems which arise in designing a personal monitor which includes
its own measuring system, most personal monitoring is done by sample collection
using the techniques described in Chapters 5 and 6.   The measurement is then made
in the laboratory using whatever technique is appropriate.

Monitoring the Environment External to the Factory

Monitoring the environment around factories involves making measurements at
locations to which access may be difficult and which may be remote from services
of electric power, water, etc.   Measurements are usually made over lengthy periods
and so the equipment often may have to operate unattended.   Further, because there
are usually a number of pollutants present at low levels of concentration, the
measuring techniques must have high selectivity and also high sensitivity at very
low concentration levels.

In view of these requirements it is very common for monitoring in the environment
to be carried out by sample collection followed by laboratory analysis.   By this
means the specialised laboratory instruments can be used to provide the sensitivity
and selectivity required.   However, provided monitoring points can be provided
with supplies of power, etc., it is possible to use modifications of these laboratory instruments in the field to make direct measurements on the atmosphere without
prior sample collection.

A monitor based on a gas chromatograph fitted with a semi-specific detector using
the technique of chemi-luminescence, has been used for measuring low concentrations
of vinyl chloride in the atmosphere. (Ref. 22).   A gas chromatograph fitted with
a sample enrichment system based on low temperature trapping is commercially available. (Ref. 23).   Optical detection systems have also been used.   A monitor
based on laser opto-acoustic spectrometry has been described (Ref. 24) and infra-red

spectrometers using long path length sampling techniques have been used. (Ref. 25).

The long path lengths required for measurement by infra-red spectrometers can be provided by means of multi-reflection folded path length cells.   The measurement made using a sample taken in a cell of this type is the average concentration at the sampling point over the time required to fill the cell, i.e. the average at a point in space over a short time.   The long path length can also be provided by removing the cell from the instrument and separating the radiation source from the detector system, or alternatively by returning the radiation from the source back to the detector by reflection from a distant mirror.   In this case the measurement is an average concentration over the absorption path length and gives no indication of the concentration at any specific point.   Interpretation of the measurement is therefore difficult and depends on assumptions made with respect to variation of concentration along the optical path.

A laser emits a beam of coherent radiation which remains parallel and does not diverge significantly over long distances.   This is a very convenient source of radiation for spectral measurements in which the beam has to travel over a long distance, because the intensity of radiation remains substantially constant over its range.   The laser source has been used in a number of applications for remote monitoring of atmospheric pollution, of which the most common involves the measurement of the radiation back-scattered by aerosols.   This technique is usually given the name LIDAR, an abbreviation of light radar, and has been used to measure the dispersion of smoke plumes from factory stacks.   A technique has been developed to measure the frequencies which arise from Raman scattering by certain molecules, such as carbon dioxide and sulphur dioxide, which can thereby be detected and determined.   The principle of laser resonance fluorescence has also been used in remote atmospheric monitoring.   In this application the frequency of a tuneable laser is adjusted to a particular transition of an atom or molecule, and the absorption of that radiation results in the radiation of another frequency characteristic of the atom or molecule.

LIDAR applications require a powerful tuneable laser source directed at the point at which measurements have to be made.   Alongside the source is a sensitive radiation detector which detects and measures the radiation which is scattered or re-radiated.   No equipment is required at the point where the pollutant exists. The time between the emission of a pulse of laser radiation and the detection of the resulting radiation by the receiver can be measured electronically, and from this information the distance of the pollutant which is detected can be determined. However, the technique has a number of limitations.   LIDAR is limited primarily to the detection of aerosols.   Again, Raman scattering cross sections are small, and very complicated equipment is required to achieve high sensitivity of measurement using this effect.   Further, resonance fluorescence is limited practically to the detection of only a few atoms and molecules.   It requires that the radiation at both the laser and fluorescence wavelengths should be transmitted through the atmosphere without substantial absorption loss.   The measurements made by these techniques approximate to measurements at a point.   They are not averages over the distance through which the beam has travelled.

The detection of pollutant gases by measurement of their infra-red spectrum using a conventional source of infra-red radiation has already been mentioned.   It is possible to replace this source by a laser source and this simplifies the optics of the instrument and virtually eliminates the loss of radiation due to divergence of the beam.   It is, therefore, possible to transmit the radiation over a much longer absorption path length than could be used with a conventional infra-red source. The instrumentation is similar to that used in LIDAR, but the radiation is reflected by some means, back to the detector which measures the intensity of the radiation received.   The difference in intensity of the transmitted and received radiation is

due to absorption over the path length.   Various systems of reflectors have been
used to return the radiation, including naturally occurring reflectors such as
aerosol clouds and foliage, and also radiation reflectors specially erected for the
purpose.   Nitrogen dioxide and carbon monoxide have been monitored using a system
of this type.   For satisfactory operation of such an instrument the laser must
emit radiation at a strong absorption frequency of the substance which is being
measured.   Modern lasers are tuneable over limited ranges of frequency but may
require refrigeration to very low temperatures, which must be controlled very
accurately to ensure stable operation of the laser.   The frequency at which
measurement is to be made must be selected to avoid interference from absorption by
other substances in the atmosphere.   Because the path length of measurement is so
great, infra-red absorptions which are weak enough to be neglected in normal infra-
red practice become significant.   The absorption of water vapour can be a serious
limitation in the choice of wavelengths at which meaningful measurements can be
made.   However, by selecting optimum operating conditions of measurement, and
using a path length of 0.61 km, carbon monoxide has been measured down to 0.005 ppm
(Ref. 26).   The measurement made by this system is an average over the absorption
path length.   The information is, therefore, subject to the same difficulties of
interpretation as that obtained from long path length measurements made by conventional spectrometers.

Laboratory Measurement Techniques

All laboratory analytical techniques have a part to play in atmospheric monitoring.
The choice of method of analysis depends on the nature of the samples, and the
range and type of substances of which the concentrations have to be determined.
In most problems those techniques which are specific to the substances being measured
are of particular value because interference from other substances which are present
is often the source of greatest error in measurement.

Standard techniques of analytical chemistry have been widely employed in atmospheric
monitoring work.   For example, a chemical reagent technique was used for determining total sulphate in the ambient air. (Refs. 27, 28).   A colorimetric method
in which iodine is liberated from potassium iodide has been used to determine
oxidants in the atmosphere. (Ref. 16) and a coulometric method for measuring the
concentration of vinyl chloride has been developed.   (Ref. 29).

However, the most commonly used analytical technique for samples taken during atmospheric monitoring is gas chromatography.   The choice of separation column and
operating conditions depends on the particular problem, and particularly on the
nature of the substances which have to be separated and measured.   The concentrations of these substances are normally very low, and so a detector which has a high
sensitivity to the substance to be measured is required.   For general purposes the
detector should be quantitative, have a response which is proportional to the
quantity present, have a wide dynamic range, and be sensitive to a wide range of
substances.   The most commonly used detector is that based on flame ionisation and
this meets the required characteristics to a high degree.

Although gas chromatography gives a measurement of the amount of a substance present,
the only information which contributes to the identification of the substance which
is being measured is the retention time of the substance on the separation column
in use.   Depending on the particular problem, this may be adequate but, in analysing
samples in which a number of components elute from the column at about the same
time, there may be difficulty in identifying the substances which give rise to the
various peaks which appear on the chromatogram.   If the substances differ in
chemical nature they may have different retention characteristics on columns of a
different type.   If a substance elutes at the appropriate retention time from two

or more columns of different composition, this provides additional confirmation of the identity of that substance.

Detectors of types other than flame ionisation can be used for special purposes. Certain detectors such as those based on flame photometry, electron capture, and alkali flame ionisation, are selective and respond only to compounds which contain certain elements. (Ref. 30). If the compound which it is required to measure contains an element to which such a detector is sensitive, the use of that detector on the gas chromatograph will allow the peak due to that substance to be recorded while ignoring peaks due to other substances which do not contain that element. This provides further confirmation of the identity of the substance concerned, and often simplifies the measurement procedure by removing from the record at least some of the chromatographic peaks. Some of these selective detectors provide lower limits of detection of those substances to which they are sensitive than is possible using the flame ionisation detector. However, the total theoretical benefit may not always be achieved because of practical difficulties associated with the use of these detectors.

When the substances concerned cannot readily be separated by gas chromatography other techniques of separation, e.g. liquid chromatography or thin layer chromatography can be used. (Ref. 31). When samples contain very large numbers of substances at low concentration it is not always possible to separate the components sufficiently well for identification and measurement by gas chromatography alone. In this case the effluent from the gas chromatograph can be flowed into a mass spectrometer where the partially separated samples are analysed for their individual components. This is a most powerful analytical technique but, because it is time consuming and requires specialised knowledge and costly equipment, it is used only in the solution of highly complex problems. For example, atmospheric samples taken at a number of points in the United States were analysed by this technique. The components of automobile exhaust were substantially resolved from each other and over twenty halogenated compounds were identified in the samples. (Ref. 32).

Techniques of optical spectroscopy have also been used in the identification and measurement of traces of pollutants in atmospheric samples. Identification of substances by infra-red spectroscopy is usually positive. Gas spectra are sharp and band shapes distinctive so that the general appearance of an absorption band, as well as its wavelength, can be used for identification. Interferences are usually obvious, and the nature of an unknown component can often be deduced from its spectrum. However, the ability of conventional spectroscopic techniques to detect very low levels of concentration is limited even with long path cells. A technique has been developed for making spectroscopic measurements using a standard spectrophotometer fitted with a long path length cell in which the pressure of the sample can be increased to 10 atmospheres. By this technique many substances can be detected and identified at concentrations far below 1 ppm. (Ref. 33).

Other techniques of optical spectroscopy have been used in atmospheric pollution studies, but these require specialised knowledge and costly equipment. A scanning interferometer with a multiple-pass long path cell, and coupled to a computer, was used to identify and measure traces of pollutants in atmospheric samples at the level of 0.001 ppm. (Ref. 34). A laser opto-acoustic spectrometer has been designed for the analysis of atmospheric samples either directly or following separation by gas chromatography. (Ref. 24).

These highly specialised techniques of measurement are of great value when complex mixtures have to be analysed in detail, when the measurement of a required substance is complicated by the presence of other substances similar in nature, and when the lowest levels of detection and identification are required. However, for the analysis of the majority of samples taken in air pollution studies a gas chromatograph suffices.   It must, however, be fitted with sampling equipment appropriate to the type of sample to be analysed, a column which will adequately separate the substance to be measured from other components in the sample, and with a sensitive detector, usually a flame ionisation device.

Few of the measurement techniques which can be used in the determination of pollutants in atmospheric samples are absolute, and almost all require to be calibrated against standard samples of known composition.   Thus the accuracy of measurement is highly dependent on the accuracy with which these standards are prepared and with which the calibration procedures are carried out.

REFERENCES

1. Organic Vapour Analyser, Century Systems Corporation, Arkansas City, Kansas, U.S.A.

2. Capabilities of Century's "G" Chromatographic Columns, Technical Note CT - 07 - 7503, Century Systems Corporation, Arkansas City, Kansas, U.S.A.

3. J.N. Driscoll & F.F. Spaziani, Trace gas analysis by photoionisation, Analysis Instrumentation Division of ISA, King of Prussia, Pennsylvania, May 1975.

4. Miran 1A Ambient Air Analyser, Foxborough Analytical Limited, Milton Keynes, Bucks, U.K.

5. Sema Electronics Limited, Irvine, Ayrshire, U.K.

6. Leakfinder, AI Industrial, Cambridge, U.K.

7. Leakseeker T.C.S. Portable Intrinsically Safe Leak Detector, AI Industrial, Cambridge, U.K.

8. Atmosphere Monitor, AI Industrial, Cambridge, U.K.

9. G.L. Baker & R.E. Reiter, Automatic systems for monitoring vinyl chloride, Amer. Ind. Hyg. Assoc. J. 38, 24 (1977)

10. Continuous Gas Monitors using Photoionisation Detection, H Nu Systems Inc., Newton Upper Falls, Mass., U.S.A.

11. Multipoint Ambient Air Monitors Protect Workers, Analysis No.1, July 1978, Foxborough Analytical Limited, Milton Keynes, Bucks, U.K.

12. IR - Gas - Monitor 675, Perkin Elmer Corp., Bodenseewerk Geratetechnik GmbH, Uberlingen, W.Germany.

13. FMS 7200 Toxic Gas Monitor, Eocom Corporation, Irvine, California, U.S.A.

14. Mass Spectrometer Pollution Monitor, G.W.B. Electronics, London, U.K.

15. Portable Atmospheric Sulphur Analyser, Meloy Laboratories Inc., Springfield, Virginia, U.S.A.

16. F. Leh, Air pollution instruments for oxidants, Industrial Laboratory, 13, Sept/Oct 1976.

17. Vinyl Chloride Monitor, Aurgesellschaft GmbH, Berlin, Germany.

18. R.G. Confer, UV – Conductivity method for determination of airborne levels of vinyl chloride, Amer. Ind. Hyg. Assoc. J. 36, 491 (1975)

19. J.A. Miller & F.X. Mueller, Evaluation of a continuous instrumental method for determinatinn of isocyanates, Amer. Ind. Hyg. Assoc. J. 36, 477 (1975)

20. Monitox Gas Detector, Compur Electronic GmbH, Munich, Germany.

21. Carbon Monoxide Pocket Dosimeter, General Electric Company, Wilmington, Mass., U.S.A.

22. W.A. McClenny, B.E. Martin, R.E. Baumgardner, R.K. Stevens & A.E. O'Keefe, Detection of vinyl chloride and related compounds by a gas chromatographic chemiluminescence technique, Environ. Sci. & Technol. 10, 810 (1976)

23. Siemens GmbH, Munich, Germany.

24. L.B. Kreuzer, Laser optoacoustic spectroscopy – A new technique of gas analysis, Anal. Chem. 46, 239A (1974)

25. C.S. Tuesday in Chemical Reactions in the Lower and Upper Atmosphere, Edit. R.D. Cadle, Interscience, New York.

26. R.T. Ku, E.D. Hinkley & J.O. Sample, Long path monitoring of atmospheric carbon monoxide with a tuneable diode laser system, Applied Optics, 14, 854 (1975).

27. P.W. West and G.C. Gaeke, Fixation of sulphur dioxide as disulphitomercurate (II) and subsequent colorimetric estimation, Anal. Chem. 28, 1816 (1956).

28. L. Thomas, V. Dharmarajan, G.L. Lundquist & P.W. West, Measurement of sulphuric acid aerosol, sulphur trioxide and the total sulphate content of the ambient air, Anal. Chem. 48, 639 (1976).

29. A. Credergren & S.A. Fredrikkson, Trace analysis for chlorinated hydrocarbons in air by quantitative combustion and coulometric chloride determination, Talanta, 23, 217 (1976).

30. D.F. Natusch & T.M. Thorpe, Element selective detectors in gas chromatography, Anal. Chem. 45, 1184A (1973).

31. R.F. Walker, D.A. Bagon, M.A. Pinches & C.J. Purnell, The Quantitative Determination of Isocyanate Components in Polyurethane Paints and their Sprays by High Pressure Liquid Chromatography and Thin Layer Chromatography. International Conference on the Monitoring of Hazardous Gases in the Working Environment, London, December 1977.

32. E.D. Pellizari, J.E. Bunch, R.E. Berkley & J. McRae, Determination of trace hazardous organic vapour pollutants in ambient atmospheres by gas chromatography /mass spectrometry/computer, Anal. Chem. 48, 803 (1976).

33. B.B. Baker, Measuring trace impurities in air by infra-red spectroscopy at 20 metres path and 10 atmospheres pressure, Amer. Ind. Hyg. Assoc. J. 35, 735 (1974).

34. P.L. Hanst, A.S. Lefohn & B.W. Gay, Detection of atmospheric pollutants at parts-per-billion levels by infra-red spectroscopy, Applied Spectroscopy, 27, 188 (1973).

Chapter 3

SOURCES OF ERROR IN SAMPLING
AND MEASUREMENT

INTRODUCTION

All monitoring systems, whether they use direct reading instruments or sample collection procedures, are subject to errors in taking the sample and in measurement.
In certain instruments the sampling system may consist of only a short length of
tube through which the sample is conveyed to the sensor. In fixed monitoring
systems, or where samples have to be obtained from locations to which the analyser,
for safety or other reasons, cannot be taken, the sampling system may consist of
lengthy tubes, filters, pumps and valves. When sample collection procedures are in
use the entire collection apparatus with which the sample comes in-contact may be
regarded as the sampling system. In all cases the sample may be changed in composition in passing through the system, though the chances of serious change are greater for these systems in which the sample residence time is greatest.

In certain sampling systems devices are incorporated which deliberately change the
state or the composition of the sample to simplify the measurement procedure. For
example, a heat exchanger may be provided to change the temperature of the sample,
substances which would interfere with the measurement may be removed by selective
reagents, or advantage may be taken of the properties of certain materials to adsorb
selectively the substances which have to be measured.

This chapter deals first with the most likely causes of error in sample handling,
whether the sample is destined for immediate measurement in a direct reading instrument or is being collected for later analysis in the laboratory. Secondly, it deals
with errors which can arise in making the measurement.

ERRORS IN SAMPLING

Distance/velocity Lag, Pressure Drop and Flow Rate in Sample Lines

In fixed monitoring systems, and also when an analyser cannot be taken into an area
where monitoring is required, it is common practice to convey the sample to the
analyser through a sample line which may be several hundred metres in length. The
sample is drawn through the line by a pump which is usually located adjacent to the
analyser. Ideally, to avoid possible contamination, the analyser should be located
between the sample line and the pump so that the sample does not pass through the
pump before the measurement is made. However, this configuration is not possible

with all types of analyser and in these cases special pumps which are unlikely to affect the sample composition should be used.

When the sample enters the sample line it takes a finite time to reach the analyser. If a measurement is made before that time elapses the result will not give a true value of the concentration of the gas which is being sampled. The delay in the sample reaching the analyser is called the distance/velocity lag, and can be calculated approximately from the volume of the sample line and the rate of flow of the sample. The volume V of a sample line of length L m and diameter d mm is

$$V = d^2/4000 \times L \text{ litres}$$

$$= 0.00078 \; d^2 \; L \text{ litres}$$

If the sample flow rate is F litres per minute the distance velocity lag is 0.00078 d^2L/F minutes. Thus if a sample flows at 10 litres per minute through a sample line 100 m in length and 10 mm in diameter the delay in arriving at the analyser will be 0.8 minutes and any measurement made before that time elapsed would be in error.

This simple calculation assumes that the flow in the sample line is uniform over its cross section. It also takes no account of any additional volume within the sampling system, such as filters which may be necessary to protect measuring instruments, flow stabilisers, etc. These devices have to be swept out by the sample before the gas leaving them is representative of the sample which is to be analysed. The time of sweep out is highly dependent on the shape of the vessel and the flow pattern within it, and can be calculated only by making assumptions on both of these parameters. It is usually preferable not to rely on calculation but to determine the distance/velocity lag by measurement. This can be done by introducing a continuous sample of known concentration into the sampling line and measuring the time taken by a high speed analyser to reach the correct reading of concentration.

The formula shows that distance/velocity lag for a line of given length can be reduced by increasing the rate of flow of the sample, or by reducing the diameter of the sample line. However, the gas is made to flow through a sample line by using a pump to create a pressure difference along the line and, in practice, the diameter of the line cannot be reduced beyond a certain limit without unduly increasing the resistance to flow, and so creating an unacceptably high pressure drop. If a sample is taken at reduced pressure at the end of a sampling line and analysed by a technique which measures the quantity of required substance (as opposed to its ratio to another substance of known concentration), the result will not be representative of the atmosphere which was sampled unless a correction is made for the reduction in pressure. This correction can be made by calculation, if the reduced pressure is measured, or by calibrating the analyser at the reduced pressure, if it can be assumed that the sampling conditions are not variable and give rise to a constant pressure drop.

The pressure drop in a plastics sampling line 12 mm in diameter has been determined experimentally. (Private communication). It was found that with lines between 100 and 500 m in length, the pressure drop was linearly dependent on the length. For a given length of line the pressure drop increased non-linearly with flow rate. However, within the range 20 to 80 litres per minute, the pressure drop did not exceed the value given by

$$D = 0.017 \; LF$$

where D is the pressure drop in millibars, for a flow rate of F litres per minute in a line L m in length.

This relation assumes that the line is free from sharp bends and terminal fittings which may substantially reduce its cross section and so increase the pressure drop. The relation should be used only as a general guide and if the value of the pressure drop across a sampling line is required to enable the correction of analytical results to be made, the pressure at the actual point of analysis should be measured.

The choice of diameter of a sample line is, therefore, a compromise between the acceptable values of distance/velocity lag, pressure drop and sample flow rate. If, as is usually the case, it is desired to limit the pressure drop and the dis- tance/velocity lag, the solution is to use a high speed pump with a sample line of moderately high diameter, say not less than 10 mm.   Such a sampling system provides a large volume of sample, much greater than is required by many analytical instru- ments.   In this case a second sampling pump must be provided to extract, from the primary sample stream, a secondary sample stream which meets the requirements of the analyser in respect of sample flow rate and pressure.

When an analyser is being used to measure, in turn, the concentrations of a sub- stance at a number of locations from which it receives samples through lines, the sampling line in use and the measurement system must be thoroughly flushed with new sample to ensure that the sample delivered to the analyser is not contaminated with a previous sample.   In certain fixed monitoring systems the change from one sample to another is made automatically by electrically or pneumatically operated valves which connect a number of sample lines in turn to a manifold from which the analyser withdraws a sample for analysis.   In such a system the delay caused by flushing the line prior to measurement can be reduced if the valves can be set so that the next sample line to be used is flushed by the pump while the previous sample is being analysed.   The sequential monitoring of atmospheric samples by devices equipped with sampling systems of this type has been described (Refs. 1, 2, 3).

Average concentrations throughout an area can be measured by two procedures using a number of sample points.   The samples may be analysed individually in sequence, or they may be mixed and the concentration of the mixture determined.   If a high concentration of the gas being monitored occurs at one sample point, the first system will record the high concentration exactly, when the sample from that point is analysed.   Some delay may be incurred, however, depending on the location of that point in the measurement cycle.   The second system will detect the leak immediately, but the sample with a high concentration will be diluted with the gases from other sample lines and so only a small change in signal will be recorded. It will not be clear whether this change is due to a single sample at high concen- tration, or to a small rise in the concentration of all samples.   Further, the accuracy of measurement of the mixed gas sampling system is highly dependent on maintaining constant and equal flows in all sample lines.

Leakage in Sampling System

While the sample is being taken the pressure in the sampling system is reduced below atmospheric pressure to some extent.   If, therefore, there are leaks in the sampling system, air will be drawn into the line at the point of leakage.   The air may not have the same concentration of pollutant as that at the point where the sample is being taken, and so could create an error in the measurement, the magni- tude of which depends on the size of the leak and the difference in composition of the atmosphere at the sampling point and the point of leakage.

Materials of Construction of Sample Lines

Sampling lines, and associated equipment such as filters, should be constructed of materials which will not react chemically with any substance likely to be present in the sample, which will withstand corrosion under the conditions of use, and on which components of the sample are unlikely to adsorb to any great extent. For example, lines made of copper should not be used to convey samples which contain acetylene. External corrosion of sampling systems can result in leaks. Internal corrosion creates a rough surface on the inside of the line, and the consequent increase in surface area tends to increase the adsorption of certain substances, leading to greater losses of the type described in the next section. Adsorption leads not only to a loss of the adsorbed substance, but can lead to loss of other substances by solution in the adsorbed material.

Adsorption Losses

Atmospheric water vapour creates many problems of adsorption in sampling systems. Fortunately the measurement of water vapour concentration is not required in toxic gas monitoring but the presence of water vapour in the sample may affect the measurement of other components. Water vapour will not emerge from a sampling line at its true concentration in the original sample until the walls of the line are in equilibrium with the water vapour concentration in the sample. Any change in the concentration of water vapour in the sample cannot be measured in the gases leaving the line until a new equilibrium has been established. This process can take several hours in cold sampling lines. The water vapour adsorbed on the walls of the sampling system can affect in a similar manner the concentrations of other substances in the sample, especially those substances which are soluble in water.

Particular care should be taken at any point where gases are cooled during sampling, for example, in a sampling line which passes outside a building. Under these conditions water may condense within the sampling lines, thereby giving rise to even greater losses of substances which are soluble in water. A remedy is to heat the sample line to ensure that condensation does not occur. If the analytical measurement is complicated by the presence of water vapour it may be advantageous to remove the water vapour from the sample, or at least to reduce its concentration. Chemically reactive drying agents such as sulphuric acid cannot normally be used because they may react with the substances which have to be measured. Certain desiccants such as calcium chloride do not retain their physical shape when they absorb water and they then obstruct the flow of sample. Molecular sieve driers retain not only the water but also remove by adsorption other substances from the sample stream. However, for any particular problem it may be possible to select a drying procedure which will remove sufficient of the water to simplify the analysis, without significantly affecting the composition of the remainder of the sample. For example, water was removed from a sample stream by a trap of magnesium perchlorate without affecting the concentration of vinyl chloride in the stream. (Ref. 4). The water vapour was reduced to a level at which a carbon adsorbent trap for vinyl chloride could be used to measure the concentration during a 24 hour exposure, whereas in the absence of the desiccant the trap showed breakthrough after about 4 hours. (See Chapter 6). (Note that magnesium perchlorate is a very reactive substance and must be handled with care).

A technique has been described for removing water vapour (and certain other specific classes of compounds) from a gas stream by diffusion of the water vapour through a membrane of a polymer of perfluorosulphonic acid (Ref. 5). The water vapour was reduced to a concentration at which it did not interfere with the measurement of traces of pollutants by long path infra-red absorption spectroscopy, using sample pressures of 10 atmospheres.

Although adsorption problems arise most often from the presence of water, any highly polar substance can be adsorbed on the surface of a sampling system. When the substance which is to be measured is highly polar the loss by adsorption from the atmospheric sample can lead to considerable error in measurement of concentration. For example, toluene di-isocyanate (TDI) is polar, exhibits low vapour pressure, and adsorbs strongly on most materials. An evaluation of materials of construction for sample lines for atmospheres containing low concentrations of TDI was undertaken (Ref. 6). The air containing traces of TDI was passed through lines of different materials at a flow rate of 0.5 litres per minute, which was the sample rate required by the analyser. When nylon lines were used the TDI was totally adsorbed within a few feet. With sample lines of polytetrafluoroethylene the loss followed classical adsorption characteristics, but lines of only two or three feet could be used before the loss became unacceptably great. However, the sample loss was directly related to the length of the line, and so it appeared that the loss at a given concentration was proportional to the surface area of the line. If, therefore, the flow rate were increased, the same quantity of TDI would be lost, but the percentage of the TDI lost from the sample would be reduced. The flow rate was, therefore, increased to 5 litres per minute. One tenth of the total flow was diverted into the analyser and the remainder of the sample by-passed the instrument. It was found that by using this system, sampling lines of polytetrafluoroethylene or glass up to 20 feet in length could be used without significant loss of TDI from the sample. Even with the by-pass system, sample lines of nylon, copper and stainless steel were unsuitable except in very short lengths. The loss of TDI by adsorption on the walls of the sampling line means that the detector does not record the true value of the concentration of TDI until the walls are in equilibrium with the sample gas. Until that time any reading taken will be in error. In the tests described, using a polytetrafluoroethylene sampling line 20 feet in length in a by-pass system, the correct reading was not obtained until 30 minutes had elapsed beyond the delay of the analytical system.

The tests were carried out using clean tubing. The effects of dirt and contaminants which may arise in industrial applications were not assessed. Further, if the flow in the sample line became turbulent, thereby increasing the contact between the TDI and the walls, the loss became greater. Laminar flow is essential for minimum loss and so the pumps should provide a smooth, not pulsating, flow, or alternatively a flow smoothing device should be included in the sampling system.

Effects of Dust in the Sampling System

Many gas sample streams contain dust which may interfere with the operation of measuring instruments. The dust is usually removed by filters, and precautions must be taken to ensure that these filters do not become so laden with dust that they obstruct the flow of the sample, or change the pressure within the sampling system. Changes due to filter blocking can most easily be observed by measurement of the pressure of the sample gas after it has passed through the filter.

However, the presence of dust within the sampling system, whether on filters or deposited inside sampling lines, can affect the composition of the sample and so give rise to errors. For example, dust has a very large surface area and it can act as a base for reaction or adsorption of some of the components of the sample. This is most likely to occur with dusts which are chemically active or which are composed of particles of carbon, especially if they contain traces of oil. A different type of error can occur when the dust particles which are trapped have been exposed to relatively high concentrations of chemicals and, in consequence, small quantities of these substances are retained, adsorbed on the surface of the dust particles. These substances are slowly released from the dust particles by the gas stream passing through the filter, and so affect the composition of the sample stream.

The possibility of such interference by dust, if present, should be tested in the
laboratory.   However, it is very difficult to ensure that any laboratory test truly
simulates conditions in practice.

Stability of the Sample

It is assumed that samples do not change in composition during the period between
the sample being taken and the time when measurement is made on it.   However,
there are a number of possibilities for change during that time which can lead to
errors.   The substance which is to be measured may be unstable under the conditions
of sampling.   Changes of composition due to this cause are most likely to occur if
some part of the sampling system has to be heated, either in the course of measurement or to avoid condensation.   Again, the substance may react with another substance present in the sampling system, for example, a material of construction, or
water adsorbed on the surfaces of the sampling system.   These two effects lead to
measured values lower than the true concentration of the substance in the sample.

A different effect can give rise to measured values greater than the true value of
the concentration in the sample.   This can occur when another component in the
sample is unstable under the conditions of sampling, and decomposes or reacts within the system to yield the substance which is being measured.   An example of this
effect arises when vinyl chloride is being monitored around plants in which it is
made from ethylene dichloride.   Samples taken around such plants usually contain
ethylene dichloride and, if these samples are heated under certain conditions in
the course of sampling, the ethylene dichloride may crack to form vinyl chloride
which was not present in the original sample.

The magnitude of the error in measured concentration which arises from any of these
causes is likely to be small, but it may be significant when very low values of
concentration are being measured.

Use of Colour Indicator Tubes and Absorbent Traps with Sample Lines

If colour indicator tubes or absorbent traps are used with sampling lines to take
samples from places which are inaccessible to the operator, the indicator tube or
trap should be attached to the end of the sample line which is inserted into the
inaccessible place, and the sampling pump should be attached to the other end.
By this means the sample flows directly into the indicator tube or adsorbent trap
without passing through the sampling line, and this procedure avoids problems which
arise from adsorption, etc.

SOURCES OF ERROR IN MEASUREMENT

Measurement Times of Direct Reading Instruments

Most devices which determine the concentration of a component in a sample by
measuring a physical property of that sample can make the measurement in a very
short time.   However, it may be necessary for the sample to be treated or processed
in some way before it is presented to the measuring device.   Again, certain
analytical devices may require a large sample to be collected or, alternatively,
may be able to accept the sample at only a very low rate.   In either case the total
measurement procedure may take some considerable time.   These processes of treatment and sample collection can cause errors in measurement and can create difficult-

ies in the interpretattion of the results.

Errors due to the time taken by the measurement process become significant only if
the concentration of the substance being measured in the atmosphere is changing
during the measurement time.   An instrument which makes a rapid measurement on a
small sample collected in a time which is short compared with the time over which
the concentration changes significantly, will follow the change in concentration.
If, however, the total time taken to collect the sample and measure the concentration is comparatively long, and during that time the concentration has changed, the
measurement made will be an average of the concentration which has existed over that
time.   Three types of measurement procedure are likely to be affected significantly
in this way.   The first includes certain detectors based on chemical reactions,
for example, colour change detectors, which require a considerable volume of sample
which must be admitted slowly to enable the chemical reaction to take place in a
quantitative manner.   Such detectors usually make a measurement over a period of
several minutes.   The second type includes certain techniques of optical spectroscopy in which the measurement is made on a large volume of sample contained in an
optical cell.   The time taken to fill the cell with the sample may be a minute or
more.   The third type includes certain electronic measuring devices which suffer
from inherent noise and in which the time constant of measurement is artificially
extended by electronic damping circuits to reduce the effect of this noise.   The
effect of noise on measurement is considered in more detail in a later section.

None of these devices can follow changes in concentration which take place within
the time of sampling and measurement, and peaks of concentration which exist for
slightly longer times are truncated.   Care should, therefore, be taken in comparing
concentrations measured by different techniques which have significantly different
measurement times, if it is suspected that the concentration is likely to have
changed during the time of measurement.   In most practical cases, however, the
difference in measurement arising from this cause is small.

Certain measuring devices require the sample to be treated in some way before the
measurement can be made.   These include certain chemical types of detectors in
which a small sample is treated chemically before measurement is possible, and also
certain detectors in which the components are separated before measurement, for
example, in gas chromatography.   In these cases the sample required is small and
can be taken rapidly, and so no significant errors arise from changes in concentration which take place during the time of sampling.   However, the treatment of the
sample results in a delay before the measurement can be made, and so the concentration recorded is that which existed at a previous time, that is, at the time when
the sample was taken.   These devices often operate on discrete aliquots of the
sample stream and are not truly continuous recorders, and their ability to follow
changes in concentration depends on the rate at which samples can be accepted for
processing and measurement.   The delay in measurement does not affect this ability
but limits the speed with which warning of change can be given.

The effects on average measurements of changes in concentration during the time of
averaging are dealt with in Chapter 9.

Interference from Other Substances

Few methods of measurement are completely specific to a single substance.   It
follows, therefore, that if a second substance is present which responds to the
measurement technique, an error will result.   The techniques most likely to be
affected in this way are those in which the measurement is based on a property the
value of which does not differ greatly in magnitude from one substance to another,
for example, thermal conductivity, refractive index, etc.   The magnitude of the

error depends on the difference in the value of the property concerned for the substance to be measured, and the values of the property for the interfering substances,
and also on their relative concentrations.   For this reason these techniques can be
used to make reliable measurements of concentration only under favourable conditions.

Certain techniques have either partial or complete specificity for the substances
which are to be measured.   For example, in infra-red absorption spectroscopic
methods it may be possible to find a wavelength at which the required substance
absorbs, but at which other components of the sample show negligible absorption.
Similarly, in mass spectroscopy, it may be possible to find a fragment that is
formed by the required substance but not by other substances in the sample.   If
these ideal situations cannot be achieved it may be possible to make corrections to
the measurements to allow for the effect of interference.   For example, if a substance is to be determined by infra-red spectroscopy, but another component in the
sample is known to be present which also absorbs at the analytical wavelength
chosen, a correction to compensate for the interfering absorption can be made as
follows.   A measurement is made at a second wavelength from which the concentration
of the interfering component is determined.   Using this value, the absorbance of
the interfering substance at the original analytical wavelength is calculated and
deducted from the measured absorbance of the sample at this wavelength.   The difference is the absorbance due to the required component, and from it the concentration of that substance can be calculated.   This correction procedure works most
satisfactorily when the wavelengths for measurement can be chosen so that one
substance absorbs strongly at the first wavelength and weakly at the second, and
the interfering substance absorbs strongly at the second wavelength and weakly at
the first.   Under these circumstances the corrections required are small and errors
introduced by the correction procedure are insignificant.   If such wavelengths
cannot be found, and measurements have to be made at wavelengths at which both substances show appreciable absorption, then methods of successive approximation or
solution of simultaneous equations have to be used.   Before such correction procedures are applied the underlying assumptions of linearity of calibration and
additivity should be checked.

Similar procedures can be adopted for other measurement techniques, for example,
mass spectrometry.

If such correction procedures cannot be employed it is necessary to resort to separation of the required substance from the interfering substances.   The technique
most commonly employed for this separation is gas chromatography, and the skill of
the gas chromatographer lies in his ability to select column packing materials and
operating conditions to make a separation sufficient for the purpose of the measurement required.   If sufficient separation cannot be achieved directly it may be
necessary to resort to selective detectors or even to coupling the gas chromatograph
to another analytical instrument such as a mass spectrometer.

Calibration

A few analytical techniques are absolute and the accuracy of measurement made by
them depends on the accuracy with which the various instrumental parameters are
known.   Most techniques of measurement, however, are relative and not absolute,
and these have to be calibrated using mixtures of known concentration before quantitative measurements of concentration can be made.   The accuracy of measurement for
these techniques depends not only on instrumental parameters, but also on the
accuracy with which such mixtures can be prepared.   This is often the factor which
limits the accuracy of measurement.

It is important to obtain calibration data which cover the whole range of

concentration over which the analytical device is to be used.   Few devices show a
response which follows a linear relation, or other simple relation, with concentration over a very wide range.   For high accuracy in measurement allowance must be
made for departure from these relations, which are usually assumed, and may be
adequate, in other analytical applications.   Many devices, however, show a response
which, having regard to other sources of error, can be accepted as linear over a
limited range of concentrations which covers those concentrations found in normal
operations.   Then it may be adequate to calibrate the device at a single value of
concentration.   This value should be chosen according to the requirements of
measurement, and may be the middle of the working range, the value of concentration
most commonly found, or some critical value, for example, the TLV of the substance
which is being measured.   It may be difficult to prepare calibration mixtures at
these low values of concentration, but the temptation to calibrate at much higher
values of concentration and to extrapolate to low values should be avoided, unless
the response relation of the analyser is accurately known, especially at very low
values.

Techniques for preparing calibration mixtures at concentrations required for atmospheric monitoring are described in Chapter 8.

Noise Level in Measurement

The output signals of most analytical instruments suffer from small fluctuations
commonly known as "noise".   These fluctuations are of small amplitude, they may
be regular or random in nature according to their origin, and the frequency may
vary from a fraction of a hertz to 100 hertz or even higher.   The origin of the
noise may be difficult to trace but the most common sources include the following.

Unsteady flow rate of sample.   If the flow of sample into a continuous analyser
is not constant, the analyser may be affected in such a way that noise appears in
the output signal.   For example, if the sample is driven into the analyser by the
action of a reciprocating pump the flow will occur for only half of each pump cycle.
The analyser may produce a record which shows the concentration at these times and
a reduced signal when the flow of sample has stopped.   The characteristic of this
type of noise is that it occurs at the same frequency as the operation of the
sampling pump, and in phase with the pulses in the flow of sample.   The noise can
be eliminated by using a different type of pump or by introducing a flow smoothing
device into the sampling system.

Mechanical vibration.   Many types of analytical instrument are sensitive to mechanical vibration which may arise either within the analytical system itself, for
example from sampling pumps, or from other equipment which is operating in the
neighbourhood.   The vibration causes slight mechanical movement within some part
of the measuring device, which gives rise to changes in the output signal and is
recorded as noise.   This type of noise, especially if it has a very low frequency,
can be difficult to remove.   However, it can usually be reduced considerably by
setting the measuring device on an anti-vibration mounting and isolating it as far
as possible from the source of vibration.   The lines which bring the sample into
the measuring device, and all other connections to it, should be flexible to avoid
the transmission of vibration along them.

Mains supply electrical noise.   Much of the noise which affects analytical instruments is electrical in origin.   If the analyser is being used in a location near
industrial electrical equipment the electrical mains supply on which it is operating

may be frequently disturbed by pulses caused by the operation of that equipment. Unless the electrical system of the analyser has been designed to operate under these conditions the pulses may disturb the measurement circuits in such a way as to give rise to noise on the analytical record.   The remedy in most cases is to provide the analyser with a smoothed and stabilised power supply.   However, certain types of industrial equipment may give rise to electrical interference which is radiated rather than transmitted along the electrical mains conductors.   This radiated interference may disturb, and create noise in certain types of analysers especially if they incorporate high-gain amplifiers.   The remedy is to shield the analyser from the interfering radiation.

Electronic noise.   Noise which is electrical in origin can also be generated within the analyser.   Most electrical devices have random noise associated with their operation, and the design of the device and its associated circuits should be such as to reduce this noise level below a value which can be tolerated in the specific application.   It is undesirable to resort to large values of electrical damping to smooth the output signal and reduce the noise level.   This practice increases the time constant of the operation of the analyser and reduces its ability to follow changes in the composition of the sample.   Electronic noise may also be generated within the measuring sensor itself.   For example, if a flame ionisation detector becomes contaminated by an extraneous organic material it will give rise to a noisy signal resulting from ions generated from that organic substance.   Such detectors must, therefore, be free from combustion products.   Generally, all analyser elements should be kept clean and correctly adjusted to reduce noise in the signals they produce.

Recording and processing of noisy signals.   The noise level recorded in an analyser signal depends greatly on the type of display or recording system in use.   Mechanical recorders introduce their own electrical and mechanical damping systems. Signals which are recorded on them may appear to have a much lower noise level, especially with respect to high frequency components, than would appear if the same signals were displayed on a system which has no inertia, for example, a cathode ray oscilloscope.   The noise signal may be regular or random depending on its origin.   Normally the noise signal will cause the instrument to record values both above and below the true value, but in certain types of devices it is impossible, for electrical or mechanical reasons, for the signal to fall below a certain value. Therefore, if the instrument is made to produce a record at that value the noise level appears to be reduced because the part of the noise signal which should appear below that value cannot be recorded.   This occurs with certain types of recorders in which a mechanical stop prevents the pen from moving below the chart zero.   To obtain a true value of the noise level it is necessary to offset the record so that the signal zero is recorded above the mechanical zero of the recorder.

If the noise signal is random in nature but the measurement signal is constant, the effect of the noise can be reduced by taking a number of discrete measurements of the total signal at intervals.   By taking an average of these values the noise components which enhance the measurement signal tend to cancel those which reduce it.   By this means the uncertainty introduced into the measurement by the noise is reduced by a factor equal to the square root of the number of measurements which are used in calculating the average.

The effect of noise level on measurement.   The noise level sets a limit to the accuracy with which a measurement may be made.   If only a single measurement is possible this cannot have an uncertainty less than that which results from the noise.   If, however, measurement is being made of a value of concentration which is not changing, and which can be averaged over a time either by smoothing techniques or statistically, the effect of noise on the accuracy of the measurement can be reduced.

The magnitude of noise from certain causes is independent of the magnitude of the signal being measured.   Then the importance of the noise level and its effect increases as the measurement value is decreased.   Thus the effect of noise level is most marked when low concentrations are being measured, and noise level is often an important factor in determining the limit of detection of a technique of measurement.   The limit of detection of a substance in a specified analytical system is the smallest amount of that substance which will produce a significant change in the output signal produced by the system.   This is dependent on the noise level within the system and on the measurement conditions employed.   A rough rule, however, is that a significant change has not taken place in the output signal until the average signal recorded has a value greater than twice the average deviation from zero due to noise.   Any signal of smaller value cannot positively be ascribed to a cause other than noise.   While this is a general guide the actual limit depends in each case on the nature of the noise signal.

Instrumental Stability

Two instrumental parameters may vary giving rise to errors in measurement.

Zero error.   The first is the zero value, which is the response indicated by the measuring system when none of the substance to be measured is present in the sample. The effect of a change in this value depends on the nature of the measurement system, but in most instruments the measured value is displaced from the true value by an amount equal to the error in the zero value.   Thus, if the deflection is measured from the zero value and not from the zero on the indicator or recorder scale, the error can be eliminated.   For this reason, if it is suspected that the zero value of the system may be changing, it is good practice to check the value by introducing a gas mixture which contains none of the component being measured, immediately before the measurement is made on the sample.   The ideal method is to introduce such a gas mixture before and after the measurement, and to read the deflection from the mean of the two values recorded.   This procedure is, however, justified only when measurements are being made to the highest accuracy, or when the zero value is changing rapidly.

Changes in zero value arise from a number of causes and their origins depend on the measurement system.   Common causes are slow mechanical changes due to small variations in temperature, changes in dimensions due, for example, to flow of plastics materials, ageing of electrical and electronic components, fouling of the analyser by components in the sample, and to changes within the sampling system, e.g. filter blocking.   The drift in the zero value may be in either direction.   For this reason, if the indicator or recorder being used has a mechanical stop so that it cannot indicate values below zero on its scale, the zero value of the measuring system should be displaced to a positive value on the recorder scale if zero drift is likely to occur.   By this means, drifts of zero value in the negative direction can be recorded.

The mixture which is used in the measuring system to record the zero value may be prepared from gases which are known to be free from the substance which is to be measured.   However, an alternative procedure is to pass an actual sample through an absorber which will remove completely this component.   If the component is an organic compound it can usually be removed by a trap of activated carbon, but precautions must be taken to avoid breakthrough.   (See Chapter 6).   If removal of the substance is incomplete the measuring system will have a zero error, which will result in low values being indicated for the concentrations of actual samples.

Sensitivity error.   The sensitivity of an instrument is the magnitude of the indication shown when a sample which contains a given concentration of the substance is being measured.   Changes in this value can arise from causes similar to those which give rise to changes in zero value.   The sensitivity of the device can best be checked by introducing a mixture which contains a known amount of the substance which is to be measured, and checking the value indicated.   For most instruments any correction which has to be applied is in the form of a factor by which the indicated value has to be multiplied to give the true value.   It is necessary to ensure that the zero value is correct before making this check.

The frequency with which these checks have to be made depends on the type of analyser, and on the conditions under which it is being used.   It is usual to check the performance of instruments frequently when they are first installed.   From the experience gained during this period suitable intervals between checks can be decided upon.   The frequency of testing zero value will not necessarily be the same as that of testing sensitivity.   Because zero error usually displaces the scale of observed values linearly throughout the entire range of the instrument, its effect is greatest when low values are being measured.   For such values the effect of a small error in sensitivity is negligible.   Under conditions, therefore, when the important values of the monitoring programme are low values recorded by small deflections of the indicator of the analyser, it is necessary to avoid zero errors, but sensitivity errors are of less importance.   Frequent zero checks, but less frequent sensitivity checks, would then be required.

Some monitoring devices which are fitted with automatic sampling systems have provision within these systems for the introduction of standard test gases, at appropriate times, to check the zero and sensitivity of the analyser.

REFERENCES

1. G.L. Baker & R.E. Reiter, Automatic systems for monitoring vinyl chloride, Amer. Ind. Hyg. Assoc. J. 38, 24 (1977).

2. M.A. Field & R.C. Moore, A Computer Controlled Multipoint Sampling and Measuring System for Measuring Gaseous Contaminants in the Atmosphere.   International Conference on the Monitoring of Hazardous Gases in the Working Environment, London, December 1977.

3. Resin plants meet OSHA's vinyl chloride targets, Oil and Gas J. 24 May 10, 83 (1976).

4. W. Thain (Edit), The Determination of Vinyl Chloride, Analytical Note AN 12. Chem. Ind. Assoc. Ltd., London, 3rd Edition (1977).

5. B.B. Baker, Measuring trace impurities in air by infra-red spectroscopy at 20 metres path length and 10 atmospheres pressure, Amer. Ind. Hyg. Assoc. J. 35, 735 (1974).

6. Adsorption of Toluene Di-isocyanate on Sample Lines.   Internal Laboratory Report, Universal Environmental Instruments Division of J & S Sieger, Poole, U.K.

Chapter 4

DETECTION SYSTEMS BASED ON COLOUR CHANGE

General

The measurement tube based on a chemical system of colour change, and commonly called detector or indicator tube, is the most generally used portable detection system for toxic gases and vapours.   It consists of a glass tube sealed at both ends, and filled with a solid granular material, such as alumina or silica gel, which has been impregnated with a reagent which changes colour in the presence of the gas or vapour whose presence is to be detected, and of which the concentration is required to be measured.   Colour indicator tubes were originally made for the rapid measurement of concentration of carbon monoxide, but tubes are now commercially available for the measurement of a wide range of gases and vapours. (Refs. 1, 2, 3, 4).   Manufacturers list over 200 different tubes for the detection of gases in various concentrations and mixtures, and these are continually being added to as new requirements arise and tubes are developed to meet them.

The operations involved in the use of indicator tubes are very simple.   The sealed ends are broken off the glass tube to allow the sample gas to come in contact with the detecting reagent.   One end of the tube is inserted into a holder in which it makes an air-tight seal.   The holder is attached to a pumping device and, on operation of this device, a specified volume of the sample is drawn through the indicator tube.   The concentration of the substance to which the tube is sensitive is then determined by one of two procedures according to the construction of the indicator tube.   In one system the entire length of the reagent changes colour by an amount which depends on the concentration of the substance to which it is sensitive.   This concentration can then be determined by matching the colour which is generated with a set of standard colours which are provided with the indicator tube.

In the second and more modern system, the reagent becomes stained over part of its length, and the length of stain produced depends on the concentration of the substance to which the reagent is sensitive.   Tubes of this type usually have concentration scales printed on them, and the concentration of the substance in the sample can be read directly from this scale.   This procedure is now more common than colour matching.

Indicator tubes provide a means of determining rapidly the concentration of a gas or vapour, which can be used by semi-skilled operators.   The user does not have

39

to handle the reagents which are sealed in the tube and which are not removed in use.   However, the apparent simplicity of the system conceals certain problems, and a lack of knowledge of the method of operation of the system, and of the chemical reactions involved, can lead to misrepresentation of the results of measurement.

Factors which Control the Accuracy of Measurement

The accuracy of measurements made by colour indicator tubes is controlled by a number of factors which involve the design of the tube, the procedures adopted in using it, and the environment in which it is used.   The following factors are of particular importance.

Tube construction.   If the tube is purchased from a commercial supplier the factors relating to construction are outside the control of the user, but these factors can give a guide to the choice of tube for a particular purpose should more than one be available.

The granular support should be uniform throughout the tube.   If fines are present they may settle on one side of the tube and this can cause channelling of the flow of the sample through the tube.   This results in a stain front which is not perpendicular to the axis of the tube, a condition which is usually referred to as "tailing".   The support should not be friable otherwise fines may be generated during the normal handling of the tube.

The support itself should ideally form a pure white background against which the colour change can be observed.   The colour produced when the tube is exposed should form a good contrast to the unexposed colour.   The reagent in which the colour is generated should be deposited uniformly on the support which should be a substance which can be obtained in a high state of purity.   The chemical reactions which are used to generate the colour take place mainly on the surface, and the rate of reaction is highly sensitive to traces of impurities on the support.   In certain cases, the operation of the detection system is as highly dependent on the properties of the support as on the chemical system which generates the colour.   The manufacture of indicator tubes must be carried out under strictly controlled conditions to produce tubes which are reliable and reproducible in operation.   The details of manufacture and composition of supports and reagents are trade secrets.

The length of stain produced in a tube can be increased by increasing the particle size of the granular support.   This, however, results in the stain front becoming diffuse and undefined.   The definition of the stain front can be improved, and the risk of channelling reduced, by decreasing the diameter of the tube.   This, however, results in a greater pressure drop across the tube when the sample gas flows, and this increases the problems of sampling.   The selection of parameters for the design of tubes is a compromise between these various factors.

Pattern of flow of sample.   The sample is normally drawn through the tube by a manually operated pump, although certain types of samplers incorporate electrically operated air pumps.   Two types of manually operated pumps are in common use, bellows pumps and piston pumps.   In both cases the design is such that the pump will draw a known volume of gas, usually 100 ml, through the indicator tube at each stroke.

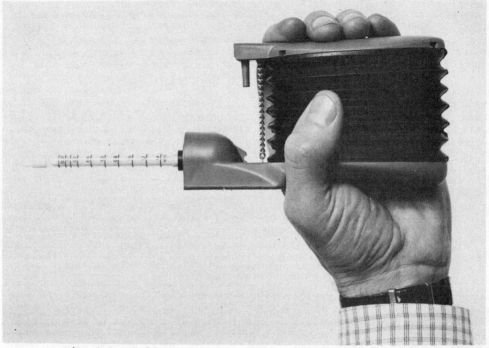

Fig. 4.1. Bellows pump.

(Courtesy of Draeger Safety)

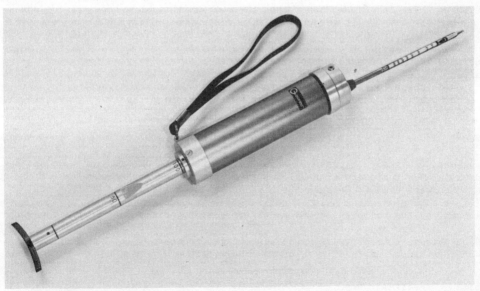

Fig. 4.2. Piston pump

(Courtesy of D.A. Pitman Ltd.)

In the bellows type of pump (Fig. 4.1) a spring retains the bellows in the fully
opened position.   The gas flow is produced by compressing the bellows and spring,
and allowing the bellows to expand again under the action of the spring.   The flow
rate is, therefore, dependent on the mechanical rate of the spring and the resistance to flow within the sample stream.   In the piston pump (Fig. 4.2) the flow is
created by reducing the pressure in the pump cylinder by pulling out the piston.
The rate of flow in a simple pump of this type is variable and depends on the manual
pressure used to move the piston, and on the resistance to flow within the sample
stream.   In some pumps of this type a limiting orifice is included in the sample
stream which limits the maximum rate of flow which can be produced by the pump.
Where no limiting orifice is incorporated in the pump the main resistance to the
sample flow is usually that of the indicator tube.   In the manufacture of tubes
for use with pumps which have no limiting orifice the parameters of tube design
which determine the resistance to the gas flow must be controlled to a high standard.

The various types of pump can be used to draw the same volume of sample through an
indicator tube.   However, the rate of flow of the sample taken by any simple manual
pump is not constant over the pump stroke.   The rate of flow produced by a piston
pump rises rapidly to a maximum and then falls slowly to zero.   If a limiting
orifice is fitted the flow rate does not rise to such a high value and remains constant over part of the stroke.

The variation of suction pressure during the pump stroke has been shown to differ
markedly from one type of pump to another and the flow rate characteristics of the
pumps have been analysed. (Ref. 5).   The effects of variation in flow pattern
within the indicator tubes are particularly important when the generation of the
stain is dependent on the pattern of sample flow.   These effects have been discussed in detail elsewhere. (Ref. 6).   Each batch of tubes is calibrated by the
manufacturer under the conditions of use in the sampling equipment which he provides.
This calibration takes account of the flow rate and flow pattern generated by the
sampling equipment specified.   Indicator tubes should not, therefore, be used with
other types of sampling pumps unless they are recalibrated under the revised conditions of sampling.   The magnitude of the errors which can arise from using an
indicator tube with a pump other than that for which it has been designed has been
reported. (Ref. 5).   It was shown, for example, that a carbon monoxide indicator
tube, designed for use with a bellows pump, indicated from 64 to 116 ppm of carbon
monoxide when used with piston pumps of various designs to test a mixture which
contained 100 ppm of carbon monoxide.   Another tube, designed for use with an
orifice-limited pump, indicated from 62 to 190 ppm on the same mixture when used
with pumps which contained no limiting orifice.   This study recommends that if
detector tubes are used with pumps of a type other than those for which they have
been calibrated, the results obtained should be regarded as only qualitative.

A recent study (Ref. 7) on the use of tubes with pumps of different types showed
clearly the effects of flow pattern on the indication by the tube.   The indicator
tubes were designed for measurement of carbon monoxide in air, and were designed
and calibrated for use with a bellows type pump, using a sample volume of 100 ml.
The tubes were used to measure the concentration of a test gas which contained 0.1
vol.% of carbon monoxide using three sampling procedures.   In the first the sample
was taken by the bellows pump using the standard operating procedures.   In the
second and third procedures the sample was taken by a piston pump in times of 10
and 30 seconds respectively.   The percentage errors shown by the measurements in
a series of three tests of a standard tube were:

Bellows pump	0%
Piston pump - 10 seconds sampling time	0% to +30%
Piston pump - 30 seconds sampling time	-20% to -40%

It is now officially recognised that the indicator tube and pump form an integrated unit.   British Standard Specification - Gas Detector Tubes - BS 5343: 1976 requires the manufacturer of the detector tubes to include in his literature a warning that tubes and pumps of different origins should not be used in combination.

In view of the importance of flow rate and flow pattern on the performance of the indicator tube, the rate of flow and the volume of sample during operation should not deviate more than 5% from that used when the tube is calibrated.

All pumps should be fitted with filters which will protect the mechanism from damage should particles of the support, or fragments of glass from the tubes, become detached during operation and be drawn into the air stream.   The limiting orifice, if fitted, must be protected in this way, otherwise, if it becomes obstructed the flow pattern of the sample will be changed.   If the orifice has to be cleaned the operation should be carried out with extreme care, because the orifice can easily be enlarged by wire reamers or by the etching action of some cleaning solutions.

Temperature and pressure.   The tubes are calibrated under standard conditions of temperature and pressure, usually 20°C and 1000 millibars.   If the conditions under which the tube is used differ significantly from the conditions of calibration certain corrections to the result of measurement may be necessary.   In the first place, the volume of gas sampled by the pump should be corrected by the normal gas laws.   It may, however, also be necessary to make corrections to compensate for changes in reaction rates with temperature.   If these changes are likely to intro- duce significant errors in measurement, correction procedures are usually supplied by the manufacturer, and should be followed closely.   Alternatively, the manufac- turer may provide information on the temperature range within which the tube will function without correction being required.

Reaction system.   Considerations of storage life limit the type of reagent systems which can be used in indicator tubes.   It is desirable from the points of view of the manufacturer, the distributor, and the user, that tubes should have a shelf life of at least two years.   This, however, cannot always be attained, especially in the early days of development of indicator tubes which contain new formulations. The shelf life of tubes is increased if they are stored at low temperatures, and it is recommended that all tubes should be stored under refrigeration, but that they should be allowed to reach ambient temperature before being used.   Tubes should not be used after their expiry date without first checking their performance and calibration.   Tubes which show any deterioration, usually by the development of a colour within the reagent system, should be discarded.

Certain tubes involve multiple reagent systems, that is, the gas which is to be measured is converted quantitatively to another gas which is more easily measured by a colour reaction.   Such tubes usually have a shorter shelf life than tubes which contain a simple reagent system.   This arises because there are more chemical stages which have to remain stable, and diffusion of reagents from one stage to another can accelerate deterioration.   Where such diffusion is likely to have a rapid and adverse effect on the tube, the stages are sometimes packed into separate tubes which have to be joined immediately before use.   This operation, of course, involves an additional possibility of error in that the system will not function correctly if the sections are joined together in the wrong order.

Other factors limit the choice of reagent systems which can be used in indicator tubes.   For example, the colour must be stable and must not fade before the tube can be read.   This is particularly important in tubes which are read by the colour matching method.   Again, in stain length tubes the stain, once formed, must be immobile and must not show any drift in the sample stream due to volatility of the reagents.   Further, moisture may have a significant effect on both the reaction by which the colour is generated and also on the physical properties of the support. The reaction systems used in indicator tubes must, therefore, be capable of allowing for the variation in moisture content which arises in atmospheric sampling.   The range of humidity which a tube will tolerate is usually quoted by the manufacturer. The effect of humidity on the reaction system must be considered when test mixtures are being prepared to evaluate the performance of a detector tube.   Certain re- action systems require the presence of water vapour for operation and this may be provided by water adsorbed on the chemicals within the tube.   This water may be removed if very dry gases are passed through the tube, and so test mixtures must be humified.

Specificity and interference.   Few practical problems in industrial atmospheric monitoring involve single contaminants.   However, most of the chemical reactions which can easily be used in indicator tubes are sensitive to a chemical type of compound and are not specific to single substances.   In consequence, if the sub- stance which the tube is designed to detect is present the tube will detect it. However, a reading on the tube does not necessarily mean that the specified substance is present, but it may indicate the presence of another substance which undergoes similar reactions with the reagents in the tube.   Thus the interpretation of the results obtained from indicator tubes demands a knowledge of all the components in the sample and also of possible reactions of these components with the indicator tube reagents.   Manufacturers must, therefore, supply information on the reagents and reactions on which the tubes are based.   Without this information the user cannot decide whether other components in the atmosphere which he is sampling are likely to interfere with the colour reaction in such a way as to enhance or reduce the measurement.   If information on possible interfering reactions is not available the tube must be recalibrated in the presence of other substances which may be present in the sample before it can be used with confidence.   It may be necessary first to carry out a survey of the atmosphere, using other analytical techniques, to determine what substances are present which might interfere with the reactions which generate the colour.

Certain substances which would interfere can be removed by incorporating a chemical trap through which the sample passes before entering the reagent system.   Thus, a tube developed for detecting vinyl chloride in air depends on the oxidation of the vinyl chloride to chlorine which is detected and measured by the colour generated when it reacts with o-tolidine. (Ref. 1).   Chlorine or hydrogen chloride would also be detected by this system and these are likely to be present in certain areas where vinyl chloride is being monitored.   However, they are both acid gases and so an alkaline trap is included in the tube to remove them from the sample.

The system used in this indicator tube for vinyl chloride illustrates the types of interferences which have to be considered before tubes are used in any partiuular application.   The alkaline trap provided in the tube will remove small concentra- tions of chlorine and hydrogen chloride.   If either is present in sufficient quan- tity to saturate the trap, it will break through to the reagents of the measurement system and enhance the reading.   Unsaturated halogenated compounds will pass through the alkaline trap and will undergo the same reactions as vinyl chloride.   They will be oxidised releasing halogen gas or vapour which will be detected by the colour- forming reagent.   Thus the measurement will be enhanced, but the quantitative effect of this interference depends on the extent of oxidation of the particular substances which are present, under the conditions of reaction in the indicator tube. Saturated halogenated compounds do not react to any significant extent.   Unsaturated hydrocarbons will pass through the alkaline trap, will be oxidised by the oxidant reagent, but chlorine will not be produced.   They will, therefore, not affect the

reading directly.   However, if they are present in large concentrations they could
consume the oxidant before the end of the measurement.   In these circumstances
vinyl chloride could pass through the reagent system without being oxidised and so
would be undetected.   Thus there is a possibility that the indicator tube could
read a value of concentration lower than the true value.

The magnitude of these interferences is quoted by the manufacturer.   Perchlore-
thylene, trichlorethylene and vinylidene chloride are indicated by the tube with
about twice the sensitivity of vinyl chloride, and 1.2 dichloroethane with about
one fiftieth the sensitivity of vinyl chloride.   If 100 ppm of ethylene or 1.3
butadiene is present in the sample, it consumes the oxidant to such an extent that
10 ppm of vinyl chloride is indicated as 1 ppm.

The interferences suffered by indicator tubes are likely to make them record values
which are greater than the true value rather than the reverse.   Measurements made
by them, therefore, tend to err in the direction of safety.   However, the possibility of interferences which reduce the reading must always be considered.   ·Only
with a complete knowledge of the reactions involved in the indicator tube, and of
the other components likely to be present in the atmosphere, is it possible to
interpret with confidence the readings obtained from these tubes.

Observational errors.   Certain tubes give their indication by the depth of colour
which has developed during exposure to the sample.   This colour is usually measured
by matching against a colour chart provided by the manufacturer.   The judgement as
to best match depends on the observer's visual acuity and colour discrimination,
and also on his experience with the tube.   Matching is also affected by lighting
conditions and can be very difficult under the illumination conditions produced by
the almost monochromatic light sources sometimes used in public and industrial
lighting.   Subdued daylight is best for colour matching and the exposure of indicator tubes to direct sunlight should be avoided.   Inexperienced observers find
difficulty in matching colours on the different background textures of the tube
support and the colour chart.

Again, the colour which is generated does not always have the same tint as that
shown on the colour chart.   A particular difficulty arises with tubes in which the
colour is formed by reactions which are slow under the conditions which exist within
the tube.   The gas which is being detected may be incompletely absorbed near the
inlet of the tube and a stain may be generated whose intensity varies along the
length of the tube.   The interpretation of such a stain calls for considerable
experience.

Most indicator tubes have been converted from a colour matching procedure to one in
which the length of the stain is measured.   In many tubes the concentration scale
for a standard sample is preprinted on the glass.   The measurement is much simpler
than the colour matching procedure and is less demanding on the operator's visual
abilities and more independent of lighting conditions.   However, experience is
again important, especially if the stain front is not perpendicular to the axis of
the tube.   In certain tubes the sensitivity has been increased by reduction of the
concentration of the reagents on the support.   This, however, reduces the intensity
of the colour which is formed and makes the stain front more diffuse, both of which
increase the difficulty of observation.   When the stain front is diffuse, ill-
defined or irregular, it is good practice to read the maximum value indicated by
any part of the stain.

In general, both colour matching and stain length tubes should be read as soon as
possible after exposure.   This reduces the possibility of error which can arise
from stain fading.   It is particularly important, however, to read colour matching

tubes at the specified time if there is any instability of the stain.

Procedures for Use of Indicator Tubes

Indicator tubes should be stored under refrigeration but should be allowed to reach ambient temperature before use.   Tubes should not be used after the expiry date of the batch unless they are rechecked and their performance is shown to be adequate for the particular application.

The tubes should be used strictly in accordance with the manufacturer's instructions, using the sampling equipment and conditions as specified by him.   The dynamic range of a measurement system based on colour measurement is usually comparatively small.   In many systems the ratio of maximum to minimum concentration which can be measured reliably, using the same sampling conditions, is only about ten to one.   The range can be extended by changing the volume of the sample taken and this is done in practice by changing the number of strokes of the pump used to draw the sample through the tube.   Manufacturers usually provide calibration data to allow the range of measurement to be increased in this way but it cannot be assumed that the range can be further increased beyond the value quoted by the manufacturers.   A tube should be used outside this range only after a full evaluation and recalibration, and calibration data should not be extrapolated beyond the range over which it has actually been established.

Tubes must be inserted in the holder in such a way as to cause the sample gas to flow in the correct direction, which is indicated by some means on the tube.   Where multiple tube systems are used the tubes must also be connected in the correct order.   The direction of gas flow through the system is particularly important in tubes which contain chemical traps, and in multi-layer tubes in which several reactions take place in sequence.   It is also important in tubes on which there is a preprinted scale of concentration.   There is, however, one exceptional circumstance in which a tube may be used with reversed sample flow.   A tube which contains a chemical trap can be used with reversed gas flow to obtain information on the presence of a substance which is removed by the trap during the normal operation of the tube.   The indication given is only qualitative unless the system has previously been calibrated.

Tubes should not, as a general practice, be reused even if they have given a zero reading on the first occasion of use.   Certain types of tubes may perform satisfactorily on a second sample, but many reagent systems are adversely affected by acid or alkaline or other gases in the atmosphere, or even water vapour.   Indicator tubes which contain chemical traps or multiple reagent systems are particularly susceptible to affects of this type.

The concentration of gas or vapour in inaccessible places can be measured using an indicator tube attached to a sampling line.   The tube should be fitted to one end of the sampling line, which should have an internal diameter approximately equal to that of the indicator tube, and the pump should be connected to the other end. The system must be free from leaks otherwise the full sample volume will not be drawn through the indicator tube, which should be placed at the point at which the measurement is required.   The pump should be operated in the normal manner except that time should be allowed between strokes to allow the pressure to return to atmospheric within the sampling tube.   Sampling tubes of up to 5 m in length can be used in this manner without creating any appreciable error in measurement.

The sampling pump should be tested from time to time according to the manufacturer's instructions.   The tests include a leakage test, usually by placing an unopened indicator tube in the holder, operating the pump and observing the time taken to reinflate the bellows (bellows pump) or for the pressure to return to atmospheric (piston pump).   The flow rate should also be checked by the method specified by the manufacturer.   The flow rate can be measured by a precision rotameter or by a bubble flowmeter of the type described in Chapter 7.   Pumps which incorporate a limiting orifice can usually be checked for correct flow rate without a detector tube in position.   Other pumps must be checked with an indicator tube in place because the resistance to flow of the tube is a factor which determines the rate of flow.   The most common reason for low flow rate is partial blockage of the filters or the limiting orifice, if fitted.

Pumps should not be used for quantitative work if they suffer from leakage or if the flow rate deviates by more than 5% from the manufacturer's specification.   The relation between rate of flow and tube response is not linear and so simple factors cannot be applied to correct for errors in flow rate.

Time Weighted Average Measurement by Indicator Tube

Long term indicator tubes.   Indicator tubes as described above are designed for grab sampling, that is, for the measurement of concentration over a short time, using a sample of perhaps a few hundred ml in volume.   There is often a need for measurement of concentration averaged over a much longer time, and special types of indicator tubes have been designed for this purpose.   It should be stressed that most tubes which have been designed for short time sampling cannot be used satisfactorily for time-weighted average measurements by reducing the flow rate to obtain the same volume of sample in a much greater time.

Indicator tubes for time-weighted average measurements must determine the average concentration regardless of fluctuations in the values of the concentration over the measuring period.   The reagent system must meet a number of requirements in addition to those of reagent systems used in tubes designed for short term measurement.   The reagent must be able to detect and measure over a very wide range of concentrations, and it must have a consistent performance throughout the whole sampling period.   This implies that it must not change due to vaporisation during that time, it must not be affected by the presence of water vapour which will collect during sampling, and it must not deteriorate as a result of being exposed to a continuous flow of air throughout the sampling period.   The rate of reaction at the beginning of the sampling time must be the same as at the end, and the indication given by the system must be a measure of the absolute amount of gas or vapour to which the tube is sensitive, and which has passed through the tube during the sampling period.   This is necessary if the time-weighted average is to be measured over different times.   Again, the rate of reaction of the gas which is being measured with the detection reagent must be much higher than the rate of other reactions which may take place on the reagent or the support.   Finally, the colour formed must be stable.   It should not fade within the sampling time to any extent which would make reading difficult or ambiguous, and, once formed, the stain must not drift along the tube, by volatility or other mechanism, even if air which does not contain the component which is being measured is flowed over it for considerable periods.

The flow of sample for time-weighted average tubes is normally provided by an electrically driven pump.   The flow rate must be controlled and maintained within a range which is defined by the nature of the reagent system.   Pumps of the types described in Chapter 7 can be used to provide samples at the flow rates required for time-weighted average measurements with indicator tubes.   If the pump is of the small personal sampler type this system becomes an attractive personal monitor.   Its particular advantage over most other types of personal monitor is that the amount of gas to which the detector has been exposed can be read at any time without any analysis being required.

Depending on the reagents contained in the indicator tube, the effluent gases from
the tube may contain corrosive substances.   These substances will pass through the
sampling pump but care should be taken to ensure that they are exhausted from the
pump in a manner which does not allow them to come in contact with the pump mechanism.

A number of tubes which meet the requirements for long term sampling and the
measurement of time-weighted average concentrations have been developed. (Refs. 1,
3).   The range of substances which can be measured in this way is currently limited
but other substances are being added as tubes are developed.   The same care should
be applied in interpreting the results from these tubes as from short term tubes.
In particular the effects of interfering substances must be considered.   As with
short term tubes the effect of interfering substances is usually to give a value
of concentration for the substance which is being monitored which is greater than
the true value.   However, again a knowledge of the reactions involved in the tube,
and of all the substances present in the sample, is required to interpret with confidence the results obtained.

Sample collection and short term indicator tube.   Time weighted average measurements can be made by collecting average samples by the methods described in Chapters
5 and 6, and making the measurement of the concentration of the contaminant in the
collected sample using an indicator tube.   In this case the measurement is made
over a very short time and so a long term tube is not required.   Short term tubes
can be used, and the time-weighted average concentration is calculated from the
result of the measurement and any concentration factor which has arisen as a result
of the method of sample collection.   If a concentration stage is incorporated in
the sampling procedure a lower limit of detection than the nominal detection limit
of the indicator tube can be achieved.

The normal precautions regarding interference must be observed unless the sampling
system includes a stage which separates the required substance from others which
would interfere with the measurement.

A device for analysing by indicator tubes substances which have been trapped on
solid adsorbents is described in Chapter 6.

PAPER TAPE DETECTOR SYSTEMS

General

Detection systems based on colour changes resulting from reaction between atmospheric pollutants and reagents held on paper tape have been used for many years.
An example is the system for detection of hydrogen sulphide by the discoloration of
paper which has been impregnated with lead acetate.   Most of the early systems
were designed to give only qualitative or semi-quantitative results but, more
recently, systems have been developed which are capable of quantitative measurement.

The basic principles of generating and measuring colour change which have already
been outlined for colour indicator tubes apply also to paper tape detector systems.
There are, however, a number of differences between the two systems which arise
from the form of the reagent support and which affect the operation and use of the
paper tape detector.

In the preparation of the paper tape the reagent which forms the colour is dissolved in a suitable solvent.   The paper is impregnated with the solution and the solvent evaporated, leaving the reagent distributed throughout the paper.   In use a known volume of sample is drawn through the paper, and the colour formed by the reaction of the reagent with the substance to which it is sensitive is measured.   The paper must, therefore, be sufficiently porous to allow the sample to pass through it, but free from holes which would allow the sample to pass through without coming in contact with the reagent.   The paper must be uniform in quality to ensure that the reagent is uniformly deposited during the evaporation of the solvent.   It is, however, rare for the paper to be as uniform in surface texture as is the granular support which is used in indicator tubes.   As a result colour matching may be more difficult than with indicator tubes because the colour may differ slightly over the area of the stain.

The paper tape is not usually so well sealed, in storage and in use, as are the contents of an indicator tube.   In consequence certain types of reagent, for example those which are affected readily by exposure to the atmosphere or which depend on the presence of moisture, cannot readily be used in paper tape systems. The reactions which are used must be capable of taking place in the short time during which the sample is in contact with the paper.   It is difficult to use multistage reactions or chemical traps unless these are placed in tubes in the sample lines, thereby complicating what is essentially a simple system.   The colour change reactions are sensitive to temperature change to the same extent as those in indicator tubes.   The measurement must be made by observing the extent of the colour change and no procedure analagous to measurement of the length of stain in indicator tubes has been developed.

The paper tape system, however, has a number of advantages over the indicator tube system, within the range of substances for which a suitable reagent system is available.   The paper detector elements are cheaper than the glass indicator tubes. The sample flow rate and volume are usually mechanically controlled thereby reducing errors which can arise from this source.   The colour is generated on a paper background which is as near as possible in texture to the comparison chart, and so colour matching is usually easier than when indicator tubes are used.   The matching is often simplified if holes are punched in the centres of the colour standards and the unknown colour is placed behind the standard.   By this method the two columns are brought immediately adjacent for comparison.   However, because the two colours are on similar backgrounds, they can readily be compared by a simple photo-electric device.   It is usual to measure the difference in reflectivity of the stained and unstained paper tape to make allowance for any background staining which has taken place on the paper tape during the sensitising process.   This can be done by a simple device consisting of a light source and a photo-electric detector which measures, in turn, the light reflected by the exposed and unexposed sections of the paper tape, thereby giving a measure of the intensity of the stain produced.   It does not measure colour, and the sensitivity of the device to stains of different colour depends on the spectral distribution of the light source and the spectral response of the detector.   The measurement made by a photometer of this type is independent of the visual matching ability of the operator and the signal is produced in electrical form and can be displayed on an indicator or recorder.

Paper Tape Devices for Spot Measurements

A device for making measurements of the concentration of certain contaminants in the atmosphere, using a small volume of sample taken over a few minutes, is commercially available. (Ref. 8).   This includes a small battery driven sampling pump, a flow control valve and flowmeter, and a timer which controls the time of operation of the pump.   The sampling conditions for any particular substance are set by the flow rate and timer.   The sensitised paper tape is in the form of a small ticket which

is clamped in a holder attached to the sampling tube.   The intensity of the stain
after exposure is measured either visually by comparison with a colour chart prepared for the substance concerned, or by a photometric device.   As with indicator
tubes, the dynamic range of the colour change system is comparatively small but the
range of measurement can be extended by changing the sampling conditions and applying correction factors.

Sensitised paper tapes are available for a range of substances.

Paper Tape Systems for Continuous Monitoring

The paper stain colour change system can readily be adapted to permit continuous
monitoring of the concentration of a substance.   If the sensitised paper tape is
made in the form of a long strip it can be moved, either continuously or intermittently, in such a manner as to expose fresh reagent to the sample at the point of
measurement.   A monitor based on this principle uses a roll of sensitised paper
tape sufficient for 168 hours of continuous operation. (Fig.4.3) (Ref. 9).

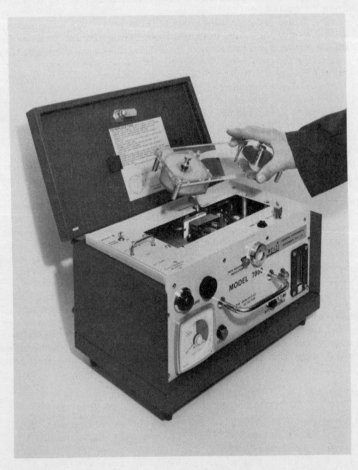

Fig. 4.3 Paper tape colorimetric analyser.
(Courtesy of J. & S. Sieger).

A small pump draws the sample at a constant and pre-determined rate through the top half of the tape and the intensity of the stain is measured after a fixed short interval by a photo-electric system. The unexposed lower half of the tape is used as a reference for the stain measurement. The paper tape is contained in a cassette which protects it from exposure to the atmosphere before it reaches the point of exposure to the sample. However, no error is introduced by a slight pre-staining of the tape, provided it is uniform over its width, because of the differential system of measurement. The fixed delay between exposure and measurement eliminates error from fading of the stain.

The dynamic range of the colour change measurement is limited and so the flow rate of the sample must be selected according to the range of concentration which is to be measured. The output indication is on a meter which has a logarithmic scale calibrated in ppm, which expands the scale at the lower values of concentration which are usually those of greatest interest. Alarms can be fitted and a recorder can be attached to give a permanent record of concentration.

Sensitised paper tape systems are currently available for the measurement of the following gases, toluene di-isocyanate, hydrogen sulphide, phosgene, nitrogen dioxide, chlorine, sulphur dioxide, vinyl chloride, hydrazine, ammonia, aliphatic and aromatic amines, formaldehyde.

The instrument which measures vinyl chloride contains within the sampling system a trap to remove chlorine and hydrogen chloride from the sample and a cartridge which converts vinyl chloride to chlorine. The chlorine is detected and measured by the standard chlorine tape and the instrument is calibrated to read directly the concentration of vinyl chloride in the sample.

Personal Monitor Based on Paper Tape

A miniature continuous monitor based on the same principle has been used as a continuous personal monitor. (Ref. 10). By using a narrower paper tape and omitting the photo-electric reading device it was possible to make a monitor small enough to be carried by a worker in the course of his normal duties. The device operates for 10 hours, and at the end of each shift the paper tape is removed. The stain pattern is scanned by a separate photo-electric reader which produces a graph showing the variation of concentration with time. Because of the delay in reading, the stain must be stable and precautions must be taken to avoid transfer of the stain between layers of the tape when it is rewound in the monitor after exposure. This is accomplished by interleaving a thin plastic tape between the layers of paper tape.

The paper tape personal monitor is similar to that based on indicator tubes in that it does not require any analysis or laboratory facilities in making the measurement. Unlike the tube device, however, it provides information only when the tape is removed and it cannot be read by the wearer during the course of the sampling period. However, when eventually the tape is examined, it provides not only a time-weighted average exposure but a complete history of the concentration to which the wearer has been exposed.

REFERENCES

1. Dragerwerke A.G., Luebeck, Germany.

2. Gastec Corporation, Tokyo, Japan.

3. Kitagawa Corporation, Japan.

4. Auergesellschaft GmbH, Berlin, Germany.

5. F.H. Coler, A study of the interchangeability of gas detector tubes and pumps, Amer. Ind. Hyg. Assoc. J. 35, 684 (1974).

6. A.L. Linch, Evaluation of Ambient Air Quality, CRC Press, Cleveland, Ohio, 1975.

7. Draeger Review, 41, 59, Draegerwerke AG, Luebeck, Germany, 1978.

8. Spot Tester, J. and S. Sieger, Poole, U.K.

9. Continuous Monitor Model 7000, J. and S. Sieger, Poole, U.K.

10. B.A. Denenberg, R.S. Kriesel & R.W. Miller. A Miniature Continuous Monitoring System for Determination of Breathing Zone and Ambient Air Concentrations of Toxic Substances. International Conference on Environmental Sensing and Assessment, Las Vegas, Nevada, U.S.A. Sept. 1975.

Chapter 5

TOTAL SAMPLE COLLECTION
METHODS

INTRODUCTION

When for any reason it is not possible to make a measurement of the concentration
of an atmospheric component by direct means on site, a sample of the atmosphere can
be collected and analysed in the laboratory.   Various means of collecting the sample
are available but the methods have different characteristics and require different
equipment.   The choice of method for any particular application depends on the
nature of the problem and the facilities available.   Two general types of sample
collecting procedures are used.   These are total sample collection, in which the
whole atmospheric sample, including the air, is collected, and selective sample
collection, in which some procedure is used to separate and retain a part of the
sample which contains the components which are to be measured.

A requirement of any sampling procedure is that sufficient sample must be collected
on which to carry out the analysis.   The amount required depends on the analytical
method which is to be used.   In a total sample collecting procedure the sample
presented for analysis is normally of the same composition as the atmosphere.   A
sensitive and selective method of analysis is therefore required and means must be
available for transferring an aliquot of the collected sample to the analyser for
analysis.   Gas chromatography is commonly used as a method of analysis.

A gas chromatograph fitted with a flame ionisation detector can detect and measure
a few nanograms of most organic substances, provided that the peak which is to be
measured can be separated from other peaks which elute at about the same time.   The
size of sample which may be injected, and in which the detectable amount of substance can be measured, depends on the nature of the column and on other factors.
The skill of the analyst lies in selecting a column and operating conditions for
the gas chromatograph which will meet the analytical requirements of each problem.

Aliquots from total gas samples can be transferred to a gas chromatograph using a
gas syringe, and may be injected directly into the column of the gas chromatograph
through an injection septum.   In this procedure, however, the sample has to be
pressurised in the syringe to overcome the pressure of the carrier gas.   If the
syringe is not absolutely leak tight at this pressure, the correct volume of sample
will not be delivered from the syringe to the column.   It is preferable, therefore,
to inject the sample at atmospheric pressure into a gas sampling valve from which
it can be transferred into the gas chromatograph.   The principle of a sampling
valve for this purpose is shown in Figure 5.1.   The sample loop should be thoroughly

purged with sample gas before the sample is transferred to the gas chromatographic column.

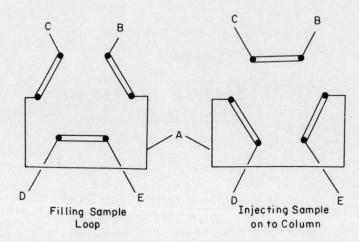

Fig. 5.1 Gas chromatography sampling valve connections.
A = gas sampling loop; B = connection for gas
syringe; C = effluent gas exit; D = carrier gas
supply; E = to gas chromatograph column.

Total samples are taken and retained in the gas phase but the physical conditions of temperature, pressure and volume of the sample may alter, according to the type of sampler, between the time when the sample is taken and the time when the measurement is made. However, provided the concentrations of the components are expressed as ratios of volume, for example, as ppm, these changes in conditions have no significant effect on the value of concentrations. However, if the concentration of a component is quoted in terms of mass per unit volume of air, the temperature and pressure of the air must be stated to ensure that the value is unambiguous. Many analytical instruments determine the mass of the substance which is being measured, although the calibration may be in terms of a ratio of volumes, using a fixed volume of sample as reference. The use of calibration data of this type will give the correct value only if the measurement is made under the same conditions of temperature and pressure as were used to determine the calibration data. If different conditions exist at the time of measurement the result must be corrected for the change in conditions, using the gas laws.

Methods of analysis, other than gas chromatography, for example, optical or mass spectrometry, may be used if these are appropriate for the substances which have to be measured. The sample handling procedures applicable to the technique of measurement must be used to transfer the sample to the analyser.

Samples should be analysed as soon as possible after collection to minimise errors which may arise from degradation and other sources. It is good practice to keep samples cool and to protect them from sunlight, both during collection and when stored awaiting analysis.

Several different types of container are used for the collection of atmospheric samples.   They differ in nature, size and means of filling and, in consequence, some are more convenient than others for particular applications.   The various techniques for collecting the samples and possible sources of errors in sampling are described below.

EVACUATED VESSEL

Grab Sampling Procedure

A sample of the atmosphere may be collected for analysis by opening a vessel, which has previously been evacuated, at the point where the sample has to be taken.   The composition of the sample is unaffected by the action of flowing into the vessel. However, the valve should be constructed of inert materials and should be of a type which does not require lubrication with·grease, otherwise certain components of the sample may be lost by reaction or solution.

The container may be made of glass or metal.   The former is less likely to give rise to problems of chemical reaction with samples, but conveying evacuated glass containers can give rise to a hazard.   All evacuated containers should be protected from rough handling to avoid implosion or collapse.   The interior walls of the vessel should be clean and free from grease to avoid loss of components of the sample by solution.   The vessel need not be completely evacuated but, if it is not, the residual air left in the container will dilute the sample, and a correction for this dilution is required.

Gas samples taken in evacuated vessels may be altered in composition by the adsorption of certain components on the walls of the vessel.   The effect is greatest with highly polar compounds which consequently may be seriously under-estimated after sampling in such a vessel.   Strongly adsorbed substances may not be removed from the walls of the vessel even when it is re-evacuated.   Instead, they may slowly desorb later and can contaminate subsequent samples.   This phenomenon is called "memory effect".

Evacuable containers have been made in many shapes and sizes.   Those made of glass are usually spherical, both for ease of construction and because this shape readily withstands the atmospheric pressure when evacuated.   Metal containers are not readily made in this shape, and are usually cylindrical, and must then be designed in such a way as to withstand the pressure of evacuation.   For a container of given shape, i.e. spherical, or cylindrical with constant ratio of length to diameter, the volume increases more rapidly with dimensions than does the surface area.   Losses in sampling due to adsorption are dependent on the area of the surface on which adsorption can occur.   In consequence, the errors in concentration arising from adsorption are likely to be more significant in small vessels than in larger vessels of the same shape.   However, there is a limit to the size of container which can be handled conveniently and safely, especially if it is made of glass.   Evacuated glass containers of two litre capacity have been used (Ref. 1) but vessels of about 200 ml capacity are more common.   A very small glass sampler based on a glass vial of capacity about 25 ml has been suggested.   The vial is sealed with a rubber cap.   It can be evacuated using a vacuum pump to which is attached a tube which terminates in a hypodermic needle which is inserted through the rubber cap.   When the vial is evacuated and the needle withdrawn the cap reseals the vial.   Sampling is carried out by inserting another needle through the cap, which again seals when the needle is withdrawn.

When it is filled, the vessel contains a sample at the atmospheric pressure and temperature which existed at the sampling point.   If ambient conditions at the place where analysis is to be made differ from those under which the sample was taken, the pressure in the vessel may no longer be atmospheric and, if the valve is opened, air may enter or sample may leave the vessel. If air enters the composition of the sample is altered.

A number of procedures can be used to remove an aliquot from the sampling vessel for analysis by gas chromatography. A simple method is to ensure that the pressure within the vessel is not below atmospheric and then to open the valve and insert through it, a long hypodermic needle which is attached to a gas syringe. On operating the plunger an aliquot is drawn into the syringe. The hypodermic needle does not make a gas tight seal within the valve and so air enters the container to replace the gas which is withdrawn. For this reason the needle should be inserted as far as possible into the sample container to ensure that the aliquot is not diluted by the incoming air. If a second sample has to be taken from the container after the induced air has mixed with the sample, a correction should be made when computing the concentration of the analysed gas, to allow for this dilution. The magnitude of the correction depends on the size of the container and the size of the aliquot extracted.

If this operation is carried out when the pressure within the sample vessel is less than atmospheric, an unknown quantity of air will enter when the valve is opened, and will dilute the sample. This can be avoided if the entry to the vessel outside the valve terminates in a tube of suitable diameter to which a septum can be fitted immediately after the sample has been taken. The hypodermic needle can then be inserted through both the septum and the open valve to withdraw the aliquot. Each aliquot removed in this way reduces the pressure within the vessel by an amount which depends on the size of the vessel and the size of the aliquot.

If the pressure within the syringe is less than atmospheric when the aliquot has been taken, air will flow into the syringe when the needle is withdrawn from the septum. Some of this air will be contained within the needle of the syringe and will be discarded if the needle is removed and the syringe is attached to a gas sampling valve as described earlier in this chapter. (Fig. 5.1). Any air which enters the barrel of the syringe is unlikely to have time to diffuse through the aliquot before it is injected into the sampling valve and will be removed with the portion of aliquot used to flush the sample loop. The reduction of pressure below atmospheric can be avoided if the syringe is filled with air before being inserted through the septum, and the air injected into the sampler and mixed. The syringe is then removed with its aliquot at atmospheric pressure, but correction to the concentration consequent upon the dilution may have to be made.

The corrections mentioned are usually small and may often be neglected. However, the possibility of errors arising from these sources should be recognised, and the corrections should be made when measurements are being made to high accuracy or under conditions of extremes of ambient temperature or pressure.

Procedures for obtaining aliquots for measurement by other analytical techniques may have to differ from those described according to the requirement of the method of measurement.

The Vacu-Sampler. A disposable partially-evacuated sampler is commercially available (Ref. 2). This consists of a partially-evacuated rigid container which has a total volume of 370 ml. During manufacture of the sampler the container is totally evacuated and then back-filled with nitrogen to a pressure of 20 inches of mercury, i.e. a volume of 246.6 ml of nitrogen at standard pressure and temperature is added to the vessel. The sampler is then sealed. To take a sample a safety cap is removed and a valve in the sampler is opened by depressing an actuator.

A sample is then drawn into the container until the internal pressure equals the atmospheric pressure, a process which takes about 10 seconds, and the sampler is resealed by releasing the actuator. The sampler is fitted with a septum through which samples can be withdrawn for analysis, or through which internal standards can be introduced to improve quantitative accuracy. The presence of nitrogen within the vessel at a pressure of two thirds atmospheric avoids most problems of losses by adsorption. However, a dilution factor of approximately three must be applied to the analysis to find the concentration of the measured components in the original atmosphere. This factor is, however, affected by the atmospheric temperatures and pressures at the sampling point and the point where the analysis is carried out, and so further correction must be made using the gas laws. The presence of the diluent gas in the sampler reduces the sensitivity of monitoring using this procedure.

Time-Weighted Average Sampling Procedure

The devices described above are fitted with inlet valves which, when opened, allow the vessel to be filled with sample in a few seconds. If, however, the flow is limited to a very low value, an evacuated sampler can take a sample continuously over a period of hours. The rate of flow of sample into an evacuated container, at any time, depends on the difference in pressure inside and outside the sampler at that time. The pressure within the vessel rises during the sampling but, in grab sampling, where the sample is taken in a few seconds, the effect of the resulting change in flow rate is insignificant. For time-weighted average sampling over long periods, specially designed inlet valves must be used to ensure that the flow rate remains constant over the sampling period. Two methods of maintaining constant rates of sampling have been described, one of which depends on the characteristics of a critical orifice and the other uses a variable flow control valve.

Critical orifice sampler. In the critical orifice sampler the sample flows into the evacuated container through a very small orifice precisely made to a specified diameter. If the orifice is small enough the sample will flow at a limiting velocity, and this flow rate will remain constant provided the pressure within the container does not exceed half the atmospheric pressure. A device which operates on this principle has been described. (Ref. 3). This consists of an evacuated container of capacity 100 ml and fitted with an orifice which has a diameter of a few μm. This device is small and light enough to make it suitable for use as a personal sampler and the flow rate is small enough to permit sampling times of 8 hours and greater if required. Evaluation of a sampler of this design has been reported. (Ref. 4).

When an orifice of this size is used to control a flow of sample in an industrial application it must be protected from dust and airborne particles, and so filters are included in the sampler, ahead of the orifice. However, with a sampler of small volume, such as that described, the rate of flow of the sample is very low and so large dust particles are rarely carried in the air stream. A second consequence of the very low flow rate is that the sample is taken from a small area immediately adjacent to the sampling probe. If it is required to take a sample from, say, a worker's breathing zone, a sampling tube must lead to the precise point. Further, the tube must be of minimum volume possible otherwise the gas collected in the sampler will be much diluted by the air originally contained within the sampling tube.

After sampling has been completed the pressure within the sampler must be measured to ensure that it does not exceed half the atmospheric pressure. If it does, the sampling flow rate will not have been constant throughout the sampling period.

The device which is used to measure the pressure should have a volume which is small compared with that of the evacuated container, and should not introduce other gas which would dilute the gas which has been collected.

The reduced pressure within the container creates certain problems in transferring aliquots from the vessel to the analyser. The sampler can readily be attached directly to the sampling system of a mass spectrometer, but if a sample is required at atmospheric pressure for other types of analysers, for example a gas chromat-ograph, an intermediate sampling stage is required. It is possible to increase the pressure within the sampler to atmospheric by the addition of a gas, for example nitrogen, which is free from the components which are to be measured. This pro-cedure, of course, dilutes the components in the gas which has been collected and increases the problems of measurement. A sample cannot readily be taken by a gas syringe against such a pressure difference and the most satisfactory procedure is to use a gas handling system which can be evacuated, and to which the sampler can be attached for pressure measurement and extraction of the sample. (Ref. 4).

There are two potential sources of error in collecting a sample by this procedure. The first is that the rate of flow of a gas through a small orifice is dependent on its molecular weight and so there could be a bias in favour of gases of low molecular weight. However, tests have been made using mixtures of hydrogen and hexane in air at low concentrations, and these showed that errors from this source are not significant. (Ref. 4). The second source of possible error arises from adsorption of chemically active or high molecular weight compounds on the walls of the sampler. Tests with mixtures of low concentrations of methane, hexane and decane in air showed that the decane was seriously under estimated. (Ref. 4). Some of the decane was recovered from the walls of the sampler by warming. A sampler has been constructed with special coatings on the interior walls to reduce adsorption of such substances (Ref. 3) but performance data are not yet available. It has been reported, however, that a sampler made of stainless steel has been used successfully to collect representative samples around an oil refinery. (Private communication).

If adsorption on the walls of the sampler becomes a limiting factor in sampling it should be possible to use the evacuated sampler instead of a pump, and to collect the organic substances on a suitable trap as is described in Chapter 6. However, the use of an evacuated sampler in this way has not yet been reported.

Flow control valve sampler. The second method of achieving a constant low rate of flow of a sample into an evacuated container is to use a flow control valve. A miniature valve for controlling the flow into a small container which is of a size which can be carried as a personal sampler has been developed. (Ref. 5). Flow rates from 0.23 ml per minute to 11.0 ml per minute are available, by regula-tion of a needle valve. The device is attached to an aluminium container which has a capacity of 123 ml and this gives sampling times from 8 hours to 10 minutes. The valve has a high pneumatic amplification factor and so the rate of flow is maintained at a constant value until the pressure within the container rises almost to atmospheric pressure.

Because the final pressure in the container after sampling is almost atmospheric, the procedure for removing a sample from the container can be simpler than that required by the critical orifice sampler. It is possible to extract a sample using a high quality leak-tight gas syringe. Otherwise the same considerations of use and possible errors of the two devices are the same.

An evacuated glass container fitted with a miniature flow control valve of this type has been used satisfactorily to sample anaesthetic gases in operating theatres. (Ref. 6).

SYRINGE SAMPLER

A sample of the atmosphere can be taken in a gas syringe by pulling out the plunger
slowly.   Errors in handling and problems of transfer for measurement are reduced
if a comparatively large sample is taken, and a syringe of capacity 100 ml and
fitted with a hypodermic needle of large diameter is recommended.   The syringe
must be free from leaks, absolutely clean, and free from lubricant.   Even small
traces of grease will dissolve certain organic substances and introduce errors in.
sampling.

Before the sample is taken the plunger should be operated to fill and empty the
syringe three or four times near the point where the sample is to be taken.   This
allows the walls of the syringe to be conditioned by gas of about the same composition as that to be sampled, and so reduces errors due to adsorption on the walls.
When a sample is to be collected the syringe should be inverted and the plunger
allowed to fall under its own weight, using manual effort only to adjust the speed
of fall.   This procedure avoids any great reduction of pressure within the syringe
during sampling, and reduces the risk of error should a leak develop.   A sample of
100 ml can be taken in this way in about 3 minutes.   After sampling the syringe is
sealed by pressing a plug of silicone rubber on to the needle, and the plunger
should be locked to avoid any movement which would change the pressure within the
syringe thereby increasing the possibility of error due to leakage.   Before taking

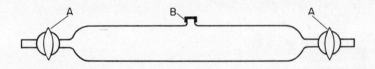

Fig. 5.2 Gas pipette sampler.   A,A = greaseless valves;
           B = septum.

an aliquot for analysis a check should be made to ensure that the pressure within
the syringe is not below atmospheric, especially if the plunger has been locked.
If there is any doubt about the pressure within the syringe slight pressure should
be exerted on the plunger to increase the pressure by a small amount.   To take an
aliquot for gas chromatography analysis the most convenient method is to remove the
needle and to attach the syringe directly to the gas sampling valve already described
earlier in this chapter.   An aliquot can then be flowed through the sample loop by
depressing the plunger of the syringe by an appropriate amount.   The procedure can
be repeated, if required, until all the sample has been removed from the syringe.

GAS PIPETTE SAMPLER

An atmospheric sample may be collected in a gas pipette (Fig. 5.2) provided a means

is available for drawing the sample through the pipette.   A water-filled aspirator
may be used, but an electrically driven sampling pump of suitable capacity is more
convenient.

The gas pipette is a glass cylinder, usually of capacity between 100 and 500 ml.
The diameter should be much smaller than the length of the cylinder to improve the
flow characteristics within the vessel and to avoid stagnant volumes.   It is fitted
with greaseless valves at each end, for example, stopcocks with barrels of polytetrafluoroethylene, and often with a rubber septum fitted near the centre, through
which aliquots may be withdrawn.   The pipette must be clean and free from grease
otherwise losses of certain components by solution in the grease may occur.

It is possible to force the sample through the pipette by a pump fitted with a
pressure outlet, but it is preferable to draw the sample into the pipette to avoid
the sample having to pass through the pump before it is collected.   A sampling
flow rate of 0.5 to 2 litres per minute is suitable and the flow should continue
until a volume of gas equal to at least twenty times the volume of the pipette has
been passed through it.   This ensures that the total volume of the pipette is
purged and that the walls become conditioned with the sample.   This reduces errors
due to loss by adsorption on the walls.   When sufficient sample has flowed through
the pipette the flow is stopped by closing first the valve on the pipette which is
adjacent to the pump, and then the valve at the other end.   This sequence of valve
closure prevents any reduction of the pressure of the sample within the pipette by
continued operation of the pump.

Samples may be held in gas pipettes provided they are chemically stable under the
conditions of storage, and do not contain components which will slowly adsorb on
the walls.   Aliquots for analysis can be removed using the rubber septum, if
fitted, or through one of the stopcocks.   The procedures, precautions and corrections for removing aliquots from a previously evacuated container described earlier
apply equally to the gas pipette sampler.

PLASTICS BAG SAMPLER

Bags made of plastics materials are commonly used for the collection of atmospheric
samples.   The bags are usually filled using a pump fitted with a pressure outlet,
but grab samples can be taken simply by pulling apart the sides of an empty bag.

Many different plastics materials have been used in the construction of sampling
bags.   Ideally, the material should be chemically inert, free from pinholes,
impermeable to gases, robust, easily sealed to make the bag, and should provide a
surface on which organic substances do not readily adsorb.   The plastics materials
available are compromises on these requirements and the most commonly used include
Tedlar (polyvinylfluoride), Saran (a copolymer of vinyl chloride and vinylidene
chloride) and Mylar (a polyester), although even polyethylene has been used when
no other material was readily available.   Gas sampling bags are easily made from
plastics film, provided means is available of sealing the film.   Bags already made
up are commercially available and are usually fitted with a metal twist-lock valve,
and sometimes with a silicone rubber septum through which aliquots can be removed
using a gas syringe and hypodermic needle.   One type of bag (Ref. 7) is made of a
material which consists of several layers of vinyl and polyester films, and is
aluminised to reduce permeation of gases.   The contents of the bag do not come in
contact with the surface on which the aluminium is deposited.

The stability of gas mixtures when stored in sampling bags has been reported. (Refs.
8, 9, 10).   Many volatile materials can be stored in bags for several hours or
even days without appreciable loss.   However, losses due to permeation can occur

with certain plastics films, and apparent losses can occur due to adsorption on the plastics surface of highly polar and less volatile components of the sample. Memory effects can arise following such adsorption and so, before they are used, bags should undergo several times a cleaning process in which they are filled with clean air, stored for several hours and pumped empty. If air free from contaminants is not available, the air used for flushing the bags should be passed through a large trap of activated charcoal. The bags should be evacuated using a mechanical pump. If this is not available the bag should be pressed flat to remove most of the air. The remaining air can be removed through the septum using a gas syringe. To avoid having to make frequent insertions through the septum it is advisable to insert a three-way valve between the needle and the syringe. By operating the valve between strokes of the syringe the system can be used as a hand pump. However, even this procedure will not remove substances which are strongly adsorbed. Desorption can be further assisted by heating the bag, but few plastics materials can withstand high temperatures. However, bags made of Tedlar can be baked at temperatures up to 80°C. A simple test for possible adsorption losses is to collect a sample in a bag and then to share the sample between the bag and another similar bag. If the contents of the two bags have the same composition it is unlikely that there has been serious loss due to adsorption. This test, however, does not check the possibility that a component has been totally removed by adsorption on the walls of the first bag.

Plastics bags should be tested for leakage each time they are used. A rapid test can be made by evacuating the bag using a suction pump through a line, in which there is a bubbler to indicate air flow between the bag and the pump. If the bag is intact the air flow will stop when the bag is evacuated. Even a small leak will be indicated by continued air flow through the bubbler. This test is not always satisfactory with bags made of the less stiff plastics materials which, on being evacuated, can collapse in such a way as to seal a leak. A more general test, which is less rapid but is suitable for bags of all materials, is to inflate the bag and store it for several hours. If the bag is free from leaks it will retain its inflated shape. Care should be taken not to over-pressurise the bag during this test.

Sampling bags should be handled with care, protected from mechanical damage, and should be shielded from direct sunlight. This can conveniently be done by placing them inside bags of strong black polyethylene.

The capacity of plastics sampling bags is usually between 0.5 and 50 litres, and the bags are filled using a constant flow of sample from a pressure pump which has a speed sufficient to fill the bag in the sampling time required. The bag then contains a sample in which the concentration of a substance is the time-weighted average concentration of that substance at the sampling point over the sampling time. If a concentration profile over the total time is required the pump can be used to deliver a sample at constant flow rate to a manifold to which are attached, through valves, a number of sampling bags. By opening and closing the valves in sequence at appropriate intervals a series of shorter term samples can be taken. If the valves are operated electrically and controlled by a suitable timing circuit the system can be made to operate without attention.

A personal monitoring system based on a plastics bag sampler has been described. (Ref. 8). The bag, which is made of aluminised polyester and has a capacity of 7.5 litres, is contained in a knapsack which is carried on the back of the worker. Also contained within a pouch of the knapsack is a small personal sampling pump which fills the bag in 8 hours. The system is cumbersome to wear but has been used to monitor the exposure of workers to vinyl chloride. With a pump which is capable of delivering a sample at a low flow rate, a sample can be collected in a bag of capacity 1 litre over a period of 8 hours. The pump and a pouch which contains the bag can both be carried on a belt which is worn around the worker's waist. (Ref. 11).

If a pump which can deliver a pressure stream of the sample is not available, or if it is considered undesirable, for any reason, that the sample should pass through the pump, an indirect method can be used to fill the bag. The bag is placed in an airtight container, and this container is evacuated slowly using a suction pump. As the pressure in the container is reduced the bag opens and fills under atmospheric pressure. The container serves to protect the bag from mechanical damage and the effects of sunlight. It is, however, advisable to have a transparent panel in the container so that the progress of filling can be observed. Difficulty has been experienced in filling bags made of the stiffer plastics materials using this procedure.

A sampling bag contains a considerable volume of sample and has usually been filled slowly. There is, therefore, a possibility that the substance which is to be measured may not be distributed uniformly throughout the volume of the bag. It is advisable to mix the contents of the bag thoroughly before withdrawing an aliquot for analysis. This can be done by "kneading" the bag for several minutes and this procedure is more effective if the bag is not filled completely during the collection of the sample. Aliquots for analysis can be removed by a gas syringe and hypodermic needle, using the silicone rubber septum, if one is fitted. Alternatively, the aliquot can be removed through the inlet tube by attaching it to the gas sampling valve of a gas chromatograph or other analyser, and applying pressure manually to the bag. The plastics bag sampler adjusts its volume automatically to maintain the contents at atmospheric pressure. Thus the pressure within the bag is unchanged when aliquots are removed.

The plastics bag sampler is particularly convenient when it is desired to use analysers which require a large volume of sample, for example long path length infra-red analysers. A large sample can be taken and almost the entire sample can easily be transferred to the cell.

REFERENCES

1. W.R. Eckert, Gas chromatographic determination of vinyl chloride monomer in polyvinyl chloride and in fat simulants, Fette Seifen Austrichmittel, 77, 319 (1975).

2. Vacu-Sampler. MDA Scientific Inc., Park Ridge, Illinois, U.S.A.

3. Critical Orifice Personal Sampler. MDA Scientific Inc., Park Ridge, Illinois, U.S.A.

4. F.W. Williams, J.P. Stone & H.G. Eaton, Personal atmospheric gas sampler using the critical orifice concept, Anal. Chem. 48, 442 (1976).

5. Vacuum Container Personal Sampler. C.F. Casella & Co Ltd., London, U.K.

6. H.T. Davenport, M.J. Halsey, B. Wardley-Smith & B.M. Wright, Measurement and reduction of occupational exposure to inhaled anaesthetics, Brit. Med. J. 2, 1219 (1976).

7. Gas Sampling Bags. Calibrated Instruments Inc., Ardsley, New York, U.S.A.

8. S.P. Levine, K.C. Hebel, J. Bolton & R.E. Kugel, Industrial analytical chemists and OSHA regulations for vinyl chloride, Anal. Chem. 47, 1075A (1975).

9. Shell Chemical Company, Analytical Method HC - 604 - 74, 1971.

10. Evaluation of a Collection and Analytical Procedure for Vinyl Chloride in Air,
    Contract No. 68 - 02 - 1408, Task Order No. 2.   U.S. Environmental Protection
    Agency, Washington, U.S.A.

11. Collection Bag and Pump System.   Sipin International Inc., New York, U.S.A.

Chapter 6

SELECTIVE SAMPLE COLLECTION
METHODS

INTRODUCTION

In selective sample collection methods only a part of the total sample is collected, but that part must include those components the concentrations of which have to be measured.   Certain systems collect all atmospheric components which have specific physical properties, for example, which condense at certain temperatures or which adsorb strongly on certain substances.   In other systems chemical traps are used to collect those components which undergo specific chemical reaction.   The first advantage of the selective sampling technique is that, because the permanent gases of the atmosphere are not collected, the volume of the sample retained is drastically reduced.   The second is that a stage of concentration is introduced which simplifies the measurement stage which follows.   However, these techniques introduce a number of sources of possible errors.   These can arise from unforeseen chemical reactions within trapping systems, the difficulty of handling the small quantities involved, and incomplete retention of the required substances in the traps.   Further, the concentration factor of the sampling procedure must be determined to enable the results of measurement to be converted to atmospheric concentrations.   This is usually done by measuring the volume of the atmospheric sample from which the selected components have been collected, and involves the use of metering pumps or other flow measurement devices.   The accuracy with which this volume can be measured is often the factor which limits the accuracy of the monitoring procedure.

The techniques of selective sampling have certain similarities to those used in the monitoring of dust.   However, in sampling particulates for hygiene measurements it is necessary to sample at a rate which ensures the collection of particles which would be inhaled during normal breathing and so a sampling flow rate of about 2 litres per minute is normally used.   Vapour and gas mixtures are, in this respect, homogeneous and the concentrations of the components are not affected by flow rate. The sample, therefore, can be collected at much lower rates.

The form in which the sample is collected and transferred to the laboratory depends on the procedure adopted for collecting the sample.   The laboratory procedures must be chosen for their ability to handle, and make the necessary measurements on, the sample in the chemical and physical form in which it is presented for analysis. The various methods of sample collection are described in the sections which follow, together with sample handling procedures appropriate to suitable analytical methods.

64

The choice of the technique of sample collection which should be used for any
particular problem must be based on the facilities available, and on the nature
and form of the pollutant which is to be monitored.   The technique of trapping on
a solid adsorbent is the most generally applicable method, but the adsorbent and
sampling conditions must be chosen specifically for each problem.   For trapping
very involatile substances a solvent scrubbing technique may be preferred, and if
these are in the form of mists a membrane filter system may be used.   The method
of collection by reaction with a specific reagent is useful where a single substance, for which there is a suitable highly specific reagent, has to be monitored,
and the condensation trapping procedure can be used effectively where a large
number of substances, which cover a very wide range of volatility, have to be
collected and measured.

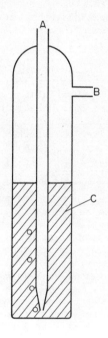

Fig. 6.1   Impinger.   A = sample entry;   B = connection to
                suction pump;   C = solvent or reagent

SOLVENT SCRUBBING

In this technique the atmospheric gases are bubbled through a solvent which will
dissolve and retain the components which have to be measured.   The trapping is
often carried out in a device called an impinger in which the solvent is contained.
This device was originally designed for the collection of dust and liquid components
of aerosols but has since been extensively applied to the collection of gases and
vapours.   Close tolerance in dimensions is required to maintain a constant flow
rate for monitoring dust.   In sampling gases and vapours, however, considerable
latitude in sampling conditions is possible without affecting the efficiency of
collection, and the dimensions of the impinger are less critical.   However, if

impingers of a standard design are used they can be changed in a sampling train
without affecting the sampling conditions.

An impinger consists of a cylindrical vessel of capacity 3 - 30 ml in which is a
central tube which is sealed into the stopper of the vessel. (Fig. 6.1).   The
solvent is contained in the vessel and the sample enters through the central tube
which terminates in a fine nozzle near the base of the vessel.   The sample is
drawn through the liquid by pumping air out of the vessel through a side arm above
the liquid level.   Impingers are usually made of glass but a less fragile type
made of plastics material is commercially available and is convenient if the device
is to be worn by an operator as a personal monitor. (Ref. 1).   The sampler should
be used in an upright position.   If there is any possibility of the sampler being
tilted to an extent where solvent might leave the container the design should include a trap to retain the solvent.   In most designs the solvent container can be
removed from the nozzle to permit solvent to be introduced or removed.   The seal
between the two parts should not be of rubber, nor should the joint be lubricated
with grease, otherwise the solvent may become contaminated.   This is particularly
important if spectroscopic analysis is to be used.

The nozzle must be designed to form bubbles small enough to present a large surface
area to the liquid.   The nozzle may be replaced by a sintered glass disc to improve bubble formation provided no particulate matter is present in the gases which
are being sampled.   The flow rate must be adjusted to a value at which discrete
bubbles are formed, and the time of contact of the bubbles with the liquid must be
as long as is necessary to dissolve the required components of the sample.   This
time can be increased by increasing the depth of solvent through which the bubbles
have to pass, or by reducing the velocity of the gas passing through the liquid.
Low gas velocities and volumes are desirable to reduce the loss of solvent by
entrainment in the gas stream.   The trapping liquid must be a good solvent for
the substance which it is required to trap and retain, and must not interfere with
the analytical method which is to be used.   The volume of the sample which is
passed through the liquid must be limited to ensure that high solvent efficiency is
maintained, and that dissolved material is not swept out by subsequent flow of
sample.   If sufficient trapping efficiency cannot be achieved in a single trap,
two or more traps can be used in series.   This effectively increases the volume
of liquid used to scrub a given volume of sample.

The efficiency of operation of a scrubbing system is affected by the vapour
pressures of the solvent and of the substance which is being collected.   Evaporation of the solvent can lead to significant losses in collected substances,
especially if these are volatile.   Collection efficiency can be improved, and loss
of solvent can be reduced, by cooling the solvent container, for example, by placing
it in an ice-bath.   The absorption efficiency is also dependent on the concentration of the component in the gas stream.   The efficiency of absorption should
exceed 90% at high concentrations, but this efficiency may fall far below this value
at very low concentrations.   In any practical case the collection efficiency of a
scrubber should be determined using two traps in series.   If more than 10% of the
required component appears in the second scrubber then a single trap is not sufficient for the particular application.

Care should be taken to ensure that liquid from the scrubber does not enter the
pump.   Further, a vapour trap should be inserted in the line between the impinger
and the pump, especially if organic solvents are being used.   A trap of activated
charcoal will remove organic vapours and a trap of silica gel will remove water
vapour.   A plug of glass wool will protect the pump against entrained liquids.

Analysis may be made by any technique which can measure the concentration of the
required substance in the solvent which has been used for trapping, in the presence

of any other substance which may have been dissolved.   Gas chromatography and
spectroscopic methods are commonly used.   The initial concentration of the component in the atmosphere is calculated from the quantity trapped and the volume of
the atmospheric gases which passed through the sampler.

TRAPPING BY SPECIFIC CHEMICAL REAGENT

The solvent in an impinger or scrubbing train can be replaced by a solution of chemicals which will react with the substance, the concentration of which is to be
measured.   The characteristics of sample collection by this method are, in general,
similar to those of trapping by solvent, with certain qualifications.   The rate of
reaction with a reagent is usually more rapid than the rate of solution in a solvent
and so shorter contact times may be employed.   Further, the product of reaction is
often less volatile than the substance which is being trapped, and so the possibility
of loss from the sampler arising from subsequent aereation is reduced.   However,
if the reaction between the sample and the reagent which is in solution causes a
precipitate to be formed, a bubbler which terminates in a sintered glass disc should
not be used because the sinter may become blocked by the precipitate forming within
it.

The reagent used in the bubbler must be chosen according to the substance with which
it is required to react, and the analytical technique must be capable of measuring
some product of the reaction which takes place.   A general reagent will react with
a wide range of atmospheric pollutants.   For example, for monitoring an acid substance, an alkaline reagent can be used which will react with all acid substances,
but the analytical technique must be capable of measuring specifically the concentration of the substance formed by the reaction of the specific acid which it is
required to monitor.   If the technique merely measures the concentration of an
ion, the possibility of this ion arising from some other source must be considered.
Thus the results of analysis must be interpreted with care.

It is preferable to use solutions of reagents of a more specific type to react with
the substance which it is required to monitor.   For example, instead of using a
general alkaline reagent, sulphur dioxide can be absorbed by aspirating the sample
through a solution of potassium tetrachloromercurate.   In the presence of sulphur
dioxide this compound forms a dichlorosulphitomercurate complex which is stable and
not oxidised by the air which passes through the solution.   The concentration can
be determined by treating the solution with para-rosanaline and formaldehyde, when
a coloured complex is generated and which can be measured by spectrophotometry.
(Ref. 2).

Most chemical trapping methods use the technique of aspirating the sample through a
solution of the reagent.   However, for certain substances the trapping reagent may
be used in a solid state.   For example, sulphur trioxide and sulphuric acid aerosol
can be trapped on filters which have been impregnated with perimidylammonium bro-
mide, resulting in the formation of perimidylammonium sulphate.   This can be
pyrolysed to produce stoichiometric amounts of sulphur dioxide which can be measured
by any convenient method.   The value of this technique is that the sulphur trioxide
and sulphuric acid react rapidly with the solid reagent and so are stabilised
against reaction with other pollutants such as metal oxides which may be trapped in
a solution. (Ref. 3).

CONDENSATION METHODS

Sample collection by condensation involves passing the atmospheric gases through a
trap, which is maintained at a low temperature, and in which organic substances

will condense and be retained.   However, the collection efficiency of a simple
trap is generally poor and depends on the chemical and physical properties of the
substances which are being trapped.   Volatile substances can be collected only if
the trap is very cold, and the low temperature must be maintained until the sample
is removed from the trap for analysis.   The operation in the field and subsequent
transport of traps at low temperature can be inconvenient.   Again, aerosols are
often formed in the trap and, unless special precautions are taken, are not retained, thereby leading to loss of sample.   Atmospheric moisture condenses in the trap,
possibly reacting with other components or creating a two phase system which complicates the analysis.   When temperatures below freezing are used for trapping,
the atmospheric water vapour in the sample may form ice and block the trap thereby
stopping the flow of sample.   If the trap has to be maintained at a very low
temperature liquid nitrogen is normally used as refrigerant.   Under these circumstances oxygen condenses within the trap and this gives rise to problems in sample
handling and analysis.   This can be avoided by using liquid oxygen as refrigerant
but this practice is usually avoided, if possible, because of the potential hazard.

A trap which overcomes most of the problems of trapping by condensation has been
described. (Ref. 4).   This trap has been used satisfactorily to collect samples of
atmospheric pollutants including aliphatic hydrocarbons from methane to nonane and
also aromatic, chlorinated and oxygenated compounds.   The trap consists of a stainless steel cylinder approximately 25 mm in diameter and 150 mm in length, one end
of which is sealed and the other end of which is threaded to take a screwed cap.
An inlet tube is welded into the cap and an exit tube is welded into the side of
the trap near the sealed end.   The exit tube runs close to the side of the trap
and terminates near the entry tube.   This layout allows the whole assembly to be
immersed in a low temperature bath.   The inlet and outlet tubes are fitted with
self-sealing couplings.   The trap is filled with stainless steel gauze rings which
improve the thermal conductivity within the trap and help to retain aerosol particles.

In use, the trap is cooled in a liquid oxygen bath, and an atmospheric sample of
50 to 100 litres is drawn through the trap by a suitable pump at a flow rate of
about 50 litres per hour.   A sampling tube made of polyethylene, 3 mm internal
diameter, can be attached to the inlet of the trap.   By this means the trap may be
located up to 60 m from the sampling point, and the liquid oxygen may be handled in
a safe area.   When sampling is completed the trap is removed from the sampling
line and pump, and the trap is sealed automatically by the valves in the connectors.
The trap is allowed to come to ambient temperature and the pollutants are then contained in the trap at a pressure of 2 or 3 bars, and concentrated about 200 times
over their concentration in the atmosphere.   The sample is transferred from the
trap by pressure equalisation into evacuated containers, and finally by Toepler
pump, until the pressure in the trap is less than 1 millibar.   Analysis is carried
out by gas chromatography or by optical or mass spectrometry.

The greatest problem arises from the presence of water vapour in the atmospheric
sample, which appears as ice in the trap.   Many pollutants dissolve in the water
after the sample is warmed to ambient temperature.   Further, the high concentration
of water complicates the analysis.   The most satisfactory method of removing the
water is to incorporate a drying agent within the trap.   Potassium carbonate is
suitable for its desiccant properties together with its ability to desorb most compounds.   It is prepared for use in the trap by being coated on the stainless steel
rings in the form of a paste, and dehydrated at 320°C before the rings are packed
into the trap.   The potassium carbonate removes some of the carbon dioxide from
the atmospheric sample and reacts with organic acids if these are present.   If such
pollutants are suspected the trap should be used without a desiccant.

Pollutants which have a concentration in the atmosphere down to a level of about

0.0001 ppm have been collected, identified and measured by this system. (Ref. 4).

COLLECTION ON A MEMBRANE FILTER

This book is concerned with the monitoring of atmospheric pollutants which are gases or vapours but, on occasions these may be accompanied by pollutants which are in the form of a mist.   For example, mercury vapour may be accompanied by a mist composed of droplets of mercury.   Similarly, gaseous hydrocarbons may be accompanied by mists of higher boiling hydrocarbons.

These mists can be collected on a membrane filter.   The material of which the filter is made, which is usually cellulose or polyvinylchloride, and the pore size of the filter, are chosen according to the nature of the mist and the substances of which it is formed.   The flow rate of the sample through the filter should be about 2 litres per minute.   At lower flow rates mist droplets, like dust particles, tend to settle and are not carried on to the filter.   Samples of about 100 litres are desirable to provide sufficient substance for subsequent handling and measurement.   However, the filter should never be allowed to become plugged with the sample material.   A possible source of error arises from evaporation of the mist particles after trapping, if they contain volatile components and the air is unsaturated with their vapour.

The substance trapped on the filter is prepared for analysis by one of two methods. A suitable solvent can be used to extract hydrocarbons and other organic substances, and analysis can be carried out by gas chromatography, optical or mass spectrometry. If inorganic materials, for example mercury mist, have been collected the filter, complete with the trapped substances, is digested in an acid which must be free from the substance which is to be determined.   Analysis in this case must be by a method appropriate to the substance which is being measured.   For metals, flame photometry, atomic absorption or X-ray fluorescence spectroscopy would be appropriate.

COLLECTION OF SAMPLES ON SOLID ADSORBENTS

Introduction

The adsorption of gases on solids is a phenomenon which has long been recognised. Scheele in 1773 and Fontana in 1777 described the removal of gases from the atmosphere by adsorption on charcoal.   The process was highly developed during World War I when traps using highly adsorbent charcoal were used for respiratory protection against war gases, and similar charcoal is now commonly used for air purification in industrial and domestic equipment.   The use of solid adsorbents to trap gases and vapours for analytical purposes has been widely reported.

The adsorption of gases and vapours on non-porous solids is extremely rapid, but the quantity retained is small because of the low surface area of the solid. Porous substances have a much greater surface area and so can retain a much greater quantity of gas.   However, if the large surface area arises from large numbers of very small pores the time taken to reach adsorption equilibrium is great, and the rate determining factor is the speed of diffusion of the gas into the pores.   When adsorption equilibrium is reached the amount of gas held per unit weight of adsorbent is a function of temperature, pressure and the nature of the adsorbed gas.

There are two main processes by which a gas is retained on a solid.   The first is physical adsorption in which the gas is held by Van der Waals' forces, and the second is chemisorption in which it is held by valence forces.   Chemisorbed

substances are, therefore, held with a greater binding energy.   Physical adsorption
is a non-specific process and arises between any solid and any gas provided the
temperature is sufficiently low.   The amount of gas which can be adsorbed, however,
depends on the nature of the substances involved.   Adsorption is rapid and takes
place as soon as the gas comes in contact with the surface of the adsorbent.
Chemisorption can take place only if there is chemical affinity between the gas and
the adsorbent.   The energy of activation must be supplied before chemisorption can
take place and, if this has a high value, the process of adsorption can be slow.
Adsorption results in a decrease in entropy for the system of the adsorbent and the
adsorbed substance.   The process is, therefore, exothermic and the heat evolved is
called the heat of adsorption.

A third process may occasionally be involved when gases are trapped on solids.   In
certain cases the vapour may enter into solid solution.   The process, however, is
very slow and the solubility of vapours and gases in most solids is very low.
Therefore, the amount of sample which is held in this way is probably insignificant
in most analytical applications.

Most of the adsorption systems used in sampling atmospheric gases depend predominantly on physical adsorption processes but, depending on the nature of the adsorbent, chemisorption processes may become involved to some extent.   The technique of
adsorption on a solid has advantages over the collection of samples in liquid
trapping agents in that the handling and transfer operations are usually less likely
to involve loss of sample.   Traps which contain solids are, in particular, more
convenient for use in personal monitoring applications.

A number of different adsorbents have been used for trapping atmospheric gases.
Silica gel has been used to collect the vapours of halogenated hydrocarbons (Ref. 5)
and for the concentration of hydrocarbons prior to analysis by gas chromatography
(Ref. 6).   It has also been used in a low temperature trap for the collection of
vinyl chloride at very low levels of concentration. (Ref. 7).   Carbon has been
very widely used as a trapping agent since the first systematic study was reported.
(Ref. 8).

Porous resins of various types have been used as adsorbents on which to collect
samples.   For example, traps which contained Tenax were used for the collection
of a wide range of substances in a general study of atmospheric pollution. (Ref. 9).

Adsorption Efficiency and Breakthrough

The capacity and the efficiency of collection of the various solids which are used
in trapping pollutants from the atmosphere are very high, but depend on a number
of chemical and physical characteristics of the substances involved, and on the
conditions under which the samples are taken.   When air which contains organic
contaminants is passed through a trap which contains an adsorbent, the organic
substances are retained within the trap initially but, as the flow continues, they
emerge from the trap with the effluent gases.   This phenomenon is known as "breakthrough", and sets a limit to the time for which the trap may be used if a true and
representative sample is required.   Beyond that time collection is no longer
quantitative.   The time to the start of breakthrough depends on certain characteristics of the adsorbent and also on sampling conditions.

Two conditions give rise to breakthrough.   In the first, the sites on which the
vapour is adsorbed become saturated.   The vapour in the incoming sample stream
cannot find vacant sites on which to adsorb and so pass through the trap with the
airstream.   The second condition arises because the vapour, after adsorption,
slowly desorbs from the solid.   It is then carried forward by the gas stream and

may be re-adsorbed farther along the trap.   Eventually, however, some vapour will
reach the end of the adsorbent bed and leave the trap.   Of these two mechanisms,
the first is of importance only when organic gases or vapours are flowing through
the trap at comparatively high concentrations, and breakthrough from this cause is
rare under the conditions of sampling used in atmospheric monitoring.   Breakthrough
during atmospheric sampling arises mainly from the second condition which is
analogous to the flow of gases through the column of a gas chromatograph.

Fig. 6.2 shows in idealised form the changes with time of the concentration of a
vapour in the effluent gases leaving an adsorbent trap.   Initially the vapour is
almost completely adsorbed in the trap but even from the outset a very low concentration will be present in the effluent.   This, however, is insignificant for most
purposes, and may be neglected.   As the adsorption front traverses the tube the

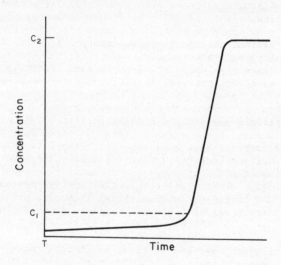

Fig. 6.2   Concentration of vapour leaving an adsorbent trap.

   C_2 = concentration of vapour in inlet gas;

   C_1 = concentration at "Breakthrough".

concentration of the vapour in the effluent rises slowly.   When the front approaches
the end of the bed of adsorbent the concentration of vapour in the effluent rises
with increasing speed until, when the front reaches the end of the bed, the effluent
reaches the same composition as the inlet gas.   Thus, breakthrough does not occur
at an instant of time, and the time of breakthrough is defined as that time at which
the concentration of vapour in the effluent reaches a specified percentage, usually
5 or 10 per cent, of the concentration in the inlet gases.   This time is, of course,
dependent on the rate of flow of the sample through the trap and a more useful concept is the "breakthrough volume" which is the volume of sample which has passed
through the trap up to the time of breakthrough.

The characteristics of a solid which determine its adsorption efficiency include the particle size, the surface area and the pore size distribution.   The obvious effect of particle size is the increase in pressure drop across a trap of given size as the particle size is reduced.   However, the particle size also determines whether mass transfer from the gas will be a factor which limits collection efficiency.   Breakthrough volume of a trap is increased as the particle size is reduced.   For any given adsorbent the surface area is related to the pore volume. The adsorption capacity of a solid is dependent on its surface area, provided that the surface is readily accessible to the gases within the contact time available. Thus, solids which have lower surface areas may be more effective as adsorbents if the surface area is composed primarily of pores of large volume into which gases may diffuse easily.

The sampling factors which determine the efficiency of collection of an adsorbent include the temperature at which adsorption takes place, the concentration of the vapour in the gas which is being sampled, the flow rate of the sample, the humidity of the sample and the geometry of the adsorbent bed.   A number of studies have been made on the efficiency of adsorption of a number of solids for specific vapours.   The results are not all readily comparable because of different conditions of experimentation and different methods of reporting results.   However, for the analyst faced with the choice of sampling conditions the following points are of importance.

Sample flow rate.   This is governed by the ability of the adsorbent to remove the vapour from the sample stream in the time of contact.   In a test of adsorption efficiency of respirator cartridges it was shown that there is no difference in adsorption capacity of carbon whether the flow rate was pulsed (to simulate breathing) or steady, and that the breakthrough times at constant concentration are proportional to the flow rate. (Ref. 10).   This implies that adsorption efficiency and capacity of carbon is independent of flow rate within the limits of the experimentation, and this result would be predicted from adsorption theory, provided none of the factors which control adsorption are rate limiting.   In practice, in atmospheric sampling this may not always be the case, and it was shown in a study of adsorption of vinyl chloride on charcoal for monitoring purposes that the breakthrough volume was dependent on the flow rate of the sample.   At lower flow rates the breakthrough volume was considerably increased. (Ref. 11).   The range of flow rates covered in the test was 0.05 to 1 litre per minute.   Modern sampling pumps can operate at these, and lower, rates reliably, and to reduce the risk of breakthrough the lowest flow rate which will give a sample of sufficient size for convenient analysis in the sampling time required, should be used.   The flow of the sample should not, however, be reduced below the rate at which back diffusion into the open end of the trap could transport to the adsorbent an amount of the substance being collected which is a significant fraction of the amount which is carried in the sample stream.   For this reason, sampling flow rates in traps of conventional design should not be less than 3 ml per minute.

Vapour concentration.   The breakthrough volume of a trap is markedly lower when it is adsorbing from a stream in which the vapour which is being adsorbed is present at high concentration.   Under these circumstances sampling conditions should be carefully chosen.   Because the concentration of vapour is high, sufficient material for analysis will be obtained from a smaller atmospheric sample than when low concentrations are being measured.   This smaller sample can be taken either in a shorter time or at a lower flow rate, according to the requirements of the measurement.   If neither a shorter sampling time nor a lower flow rate is practicable it may be necessary to enlarge the size of the trap to increase the quantity of adsorbent.

Temperature.   The breakthrough volume of a trap is reduced as the temperature is
increased.   If only physical adsorption processes are involved, for example when
gases are adsorbed on charcoal, the logarithm of the breakthrough volume is inversely proportional to the absolute temperature, as is predicted by the theory of
adsorption.   Deviation from linearity occurs if chemisorption is involved as, for
example, when polar substances are adsorbed on resins.   Therefore, breakthrough
volume can be increased by operating the trap under cool conditions.   Traps should
not be exposed to direct sunlight, and if the absorptive capacity is insufficient
at ambient temperature the trap may be cooled to lower temperatures.   Sampling at
temperatures below 0°C can introduce other problems, such as difficulties in handling of traps and blocking due to the presence of ice, but has been used satisfactorily to determine the presence of very low levels of vinyl chloride in the atmosphere. (Ref. 7).

Humidity of sample.   Breakthrough volume is reduced by the presence of moisture in
the sample.   In the collection of toluene on coconut shell charcoal the breakthrough volume was reduced by 50% when the relative humidity was increased from 0
to 80% at a flow rate of 1 litre per minute. (Ref. 12).   The extent of reduction
depends on the other parameters of sampling.   When the flow rate is reduced to the
minimum practicable the effect of humidity is also minimised.

If after selecting the optimum sampling conditions breakthrough still occurs due to
high humidity of the sample, it may be necessary to reduce the moisture content of
the sample before it enters the trap.   This practice should be avoided where possible because there is always the possibility that the desiccant will remove components of the sample other than water.   The desiccant should be chosen according
to the type of sample which is being taken, and the factors given in Chapter 3 on
the use of desiccants in sampling systems should be considered.   A desiccant trap
of magnesium perchlorate has been used succesfully with a charcoal trap in collecting samples of vinyl chloride over periods of 24 hours. (Ref. 13).

Geometry of the trap.   The trap must contain enough adsorbent to retain the amount
of substance which it is intended to remove from the sample stream, with sufficient
spare capacity to ensure that unforeseen deviations from the planned sampling conditions do not result in breakthrough.   The dimensions of the adsorbent bed are
not critical, but a long narrow bed gives rise to a high linear gas velocity over
the adsorbent and requires a pump which can deal with the pressure drop which arises
across the trap.   The same volume of adsorbent arranged in a short bed would have
a greater cross sectional area.   Such a bed would have a smaller pressure drop at
the same sample flow rate and the linear flow rate over the adsorbent would be
lower.   This can improve trapping efficiency under certain conditions, provided
the flow of sample can be distributed over the cross-sectional area of the bed.

Characteristics of Various Adsorbents

The adsorbents most commonly used are charcoal, silica gel and porous polymers of
various types.   A list of adsorbents suitable for collecting from the atmosphere
the substances which appear in the American official list of Threshold Limit Values
for 1970 has been published. (Ref. 14).

Charcoal.   It was shown in 1964 (Ref. 15) that organic vapours could be adsorbed
on activated charcoal and could subsequently be desorbed, and in 1970 an optimised
method, based on this system, for analysing certain solvent vapours in the atmosphere
was described.(Ref. 8).   Since then, traps filled with activated charcoal have been

used for collecting a wide range of organic substances from the atmosphere.
Charcoal is a non-specific adsorbent and will collect all components of a mixture.
The permanent gases oxygen, nitrogen, hydrogen, carbon monoxide and methane are
not adsorbed on charcoal at ordinary temperatures.   Ethylene, formaldehyde,
ammonia, and other gases which have a boiling point between $-100^{\circ}C$ and $0^{\circ}C$ are
partially adsorbed and those which have a boiling point greater than $0^{\circ}C$ are
readily adsorbed.   In addition, mercury vapour can be efficiently collected on
charcoal.   Water vapour is not efficiently trapped on charcoal but it reduces the
efficiency of the charcoal as a trap for other substances.

The physical properties of charcoal depend on the source and previous treatment
which it has received.   Two common sources are coconut shells and petroleum products.   Of these the former gives a more active adsorbent and is favoured by many
workers.   It has been shown (Ref. 11) that this type of charcoal traps vinyl
chloride more efficiently than charcoal from a petroleum source, when used under
the same sampling conditions.   Other types of charcoal, most of which are based
on carbonisation of synthetic polymers, show very high efficiency for trapping
organic compounds, but these charcoals are not so readily available in a form
convenient for use in traps as are the other charcoals.   Because of the differences
in the physical properties of the various types of charcoal, they should not be
interchanged in a monitoring system unless tests are made to ensure that the
efficiency of trapping has not been reduced.

Activated charcoal suitable for adsorption of gases is commercially available in a
variety of mesh sizes.   The charcoal should be prepared for sample collection by
removing all substances which have become adsorbed on it during storage, and this
is most conveniently done by heating the charcoal in a stream of nitrogen at a
temperature of $120^{\circ}C$ for at least 16 hours, or at a higher temperature up to $600^{\circ}C$
for a shorter time.   The charcoal should then be allowed to cool in a desiccator
and packed into the sampling tubes.   Alternatively, sealed trapping tubes which
have been packed with the charcoal may be purchased.   Various types of tubes which
are commercially available are described in a later section.

Adsorbed substances can be removed for analysis from the charcoal either by solvent
or by thermal displacement.   Charcoal which has been exposed to solvent is normally
discarded after use, but charcoal which has been subjected to a thermal desorption
procedure can be used repeatedly.

Silica gel.   Silica gel does not have great adsorption capacity for as wide a
range of organic substances as does activated charcoal.   It will collect organic
compounds which have a minimum of three carbon atoms, but the efficiency of collecting lower hydrocarbons from the atmosphere is poor, even when the trapping is
carried out at the temperature of solid carbon dioxide.   Silica gel adsorbs readily
compounds which contain hydroxyl groups, and many of the more common halogenated
hydrocarbons and polar organic compounds have been determined quantitatively after
trapping on silica gel.   Compounds of high molecular weight are retained more
efficiently than those of low molecular weight.

Silica gel has a greater affinity for water than for any other substance.   In
consequence the efficiency of adsorbing other compounds is much reduced when trapping from samples under conditions of high humidity.   The use of a drying trap to
reduce the humidity of the sample and so improve the efficiency of collection of
other compounds has been described. (Ref. 7).

The silica gel should be heated in a stream of nitrogen to a temperature of $400^{\circ}C$
for a period of at least 16 hours to remove adsorbed substances.   Thereafter it
should be allowed to cool in a desiccator.   The gel can be used in traps of

straight glass tubes, and traps made in this form are commercially available.   If
the trap is to be cooled during use it is convenient to make it in the shape of a
U-tube in which form it can be placed in a cooling bath.   If a tube has not been
cooled during trapping it may be sealed and stored for several days without loss
of the adsorbed material.

The adsorbed substances may be removed from silica gel by thermal desorption
methods or by solvent.   Silica gel which has undergone thermal desorption may be
re-used, but gel which has undergone desorption by solvent should be discarded.

Porous polymers.   Resins of various compositions which are suitable for use as
adsorbents are available commercially.   The most common have structures such as
polystyrene cross linked with divinylbenzene, polyvinylpyrollidine, diphenylphenyl-
ene oxide, polyaromatics and cross linked acrylic esters.   These various substances
offer a wide range of physical properties, and so the various resins have different
polarities and different characteristics of adsorption.   They generally have a low
affinity for water.   They are also moderately stable thermally and can be heated
without degradation to temperatures high enough to remove adsorbed gases.   The
physical properties which determine their characteristics of retention, which
include surface area, pore volume, pore size distribution, and polarity, are published by the manufacturers and can be used in the selection of adsorbents for
particular applications.   Selected properties of a number of resins are shown in
Table 6.1.

Pore size distribution in certain types of polymers shows very large numbers of
small pores.   Such polymers should not be used for the collection of organic substances which have large molecules to which much of the surface area will not be
accessible.   Volatile substances are less strongly retained on porous polymers
than are involatile substances.   There is a tendency, therefore, for volatile
substances to be displaced by involatile compounds and so collection favours
compounds of high molecular weight.

Most porous polymers have little affinity for water but, nevertheless, water can
be adsorbed on the polymer surfaces.   This can substantially reduce the efficiency
of trapping of other substances.

The resins are organic chemical compounds and so it is possible for them to react
with chemically reactive components in the sample.   If they do, the composition
of the resin and its physical properties may be changed.   In particular, resins
which are based on polystyrene can suffer from oxidation at temperatures above
$250^{\circ}C$, and some polymers of this type degrade at much lower temperatures.

Within the limits mentioned above the porous polymers have reasonable adsorption
efficiency for collecting organic compounds.   However, not all organic compounds
can be easily removed from the resins by thermal methods prior to analysis.   Polar
compounds such as alcohols, carboxylic acids, glycols and amines, are particularly
difficult to remove from certain polymers by thermal displacement, and have to be
removed by solvent.   The solvent must not react with the polymer but must have a
good solvent power, otherwise it will not remove the adsorbed substances quantitatively.   On occasions it may be necessary to resort to continuous extraction
procedures.

TABLE 6.1 Physical Properties of Selected Porous Polymers

Type	Supplier	Composition	Surface Area m² per g.	Pore Diameter nm	Limiting Temperature °C
Poropak P	Waters Associates	Styrene – Divinylbenzene	110	15	250
Poropak Q	"	Ethylvinylbenzene – Divinylbenzene	840	7·5	250
Chromosorb 101	Johns-Manville Products Corp.	Styrene – Divinylbenzene	30 – 40	300 – 400	250
Chromosorb 102	"	"	300 – 400	8·5	250
Tenax GC	Enka N.V.	Polyphenylene oxide	25	4	450

Construction and Use of Adsorbent Traps

Tubes which contain solid absorbents for trapping atmospheric gases are commercially available, and three systems are described later in some detail. However, certain workers prefer to construct their own traps. In making trapping tubes the following points must be considered.

The quantity of adsorbent must be sufficient to retain the amount of substance which is required for the subsequent analysis. The geometrical dimensions of the tube, and the particle size and quantity of the adsorbent, must combine to make a trap through which the sample can be drawn at the required flow rate by the pump which is available. Some means of measuring the total sample drawn. through the tube must be provided.

The adsorbent should be selected according to the nature of the substances which are to be collected. The previous section describes the characteristics of adsorbents from the point of view of trapping capacity. Certain other factors must, however, be considered. If at any time the trap is to be heated the adsorbent must be thermally stable at the highest temperature which will be reached. Further, the possibility of the sample cracking at that temperature, either thermally or catalytically because of the presence of the adsorbent or impurities within it, must be considered.

The quantity of adsorbent should be sufficient to provide excess capacity for the substance which is being collected. If there is any doubt about the capacity of the adsorbent, and breakthrough is considered a possibility, an additional small bed of adsorbent can be provided to follow the trap. If the two beds of adsorbent are analysed separately, the presence or absence of substances on the second bed will indicate whether or not breakthrough has occurred. The adsorbent may be packed in tubes of metal or of glass. Glass is preferred for tubes which have two or more beds of adsorbent because the various layers can be seen during manipulation. Metal tubes are preferred for improved thermal conductivity when the trap has to be heated during desorption. The adsorbent should be free from fines which would fill the vapour space within the trap and restrict the gas flow. It must be activated before use to remove substances already adsorbed on its surface, and to improve its adsorption efficiency. This is done by heating to an appropriate temperature in a stream of inert gas. The adsorbent is then cooled in a desiccator and stored there until required. The period of storage should be limited. The physical properties of substances used as adsorbents vary from one batch to another although the substances themselves are nominally the same. All the traps for a specific purpose should, therefore, be made from one batch of adsorbent. If it is necessary to use a different batch tests should be carried out to ensure that the performance of the new batch is satisfactory in the particular application involved.

The adsorbent should be packed in the tubes and retained in place with plugs of glass wool or plastics foam. Channelling should be avoided, both by selecting particle sizes which are small compared with the diameter of the tubes, and by exerting light pressure on the adsorbent during packing. Whenever possible, tubes should be stored and used in an upright position. If they are stored flat and settlement of the contents takes place, a channel will be formed at the top of the tube and some of the sample gas will by-pass the adsorbent. Tubes should be sealed with an air-tight cap at both ends after packing. These seals should be removed for sampling and replaced immediately afterwards. Tubes should be packed carefully to ensure that the pressure drop across them is reproducible. Then sampling conditions will remain constant even when the sampling tube is changed. Tubes which contain two beds of adsorbent which have to be analysed separately should be stored under refrigeration after use. This is particularly important if the sample contains substances which are lightly adsorbed and are prone to

diffusion.   If two tubes are used, instead of a single tube with two beds, to
check for breakthrough, they should be separated and sealed individually immediately
after sampling to avoid errors arising from diffusion which may occur later.

Desorption Techniques

The adsorbed substances have to be desorbed for measurement, and this is usually
carried out by one of two methods, solvent leaching or thermal displacement.   The
method chosen depends on the nature of the adsorbed substances and the type of
analysis which it is intended to use.   If the analysis is to be done by chemical
methods, or in solution by spectroscopy, it is convenient to remove the adsorbed
substances by a suitable solvent.   Either technique can be used to recover substances which are to be analysed by gas chromatography, though the solvent method
introduces one extra substance which has to be separated within the chromatograph.
Where analysis is to be made by spectroscopy, including mass spectrometry, in the
gas phase, thermal displacement methods are preferred.

A method of desorbing substances from charcoal by a headspace analysis technique
has been used and is described briefly later.

Solvent desorption.   For desorption of substances by solvent leaching the adsorbent
is mixed with a suitable solvent which extracts the organic substances which had
been adsorbed.   The choice of solvent is important.   It must be sufficiently active to displace the adsorbed substances, it must be a good solvent for them, and
it must not interfere with the method of analysis.   Solvents suitable for recovering from the recommended adsorbents, the substances contained in the TLV list for
1970 have been included in the table of recommended adsorbents (Ref. 14).   In
certain circumstances mixtures of solvents have to be used to obtain the correct
degree of polarity and solvent activity.   When spectroscopic analysis is being
used, the solvent chosen should be free from spectral absorption, especially in the
spectral region which is to be used for analysis.   If analysis is to be made by
gas chromatography, the solvent chosen should elute at a time well separated from
the substances which have to be measured, and preferably after them, so that the
peaks can be measured against a horizontal base line, and not against the sloping
background caused by the tail of the solvent peak.   It should, however, elute soon
after the substances of interest otherwise there will be considerable delay in
waiting for the solvent to elute before the next sample can be injected.   This
delay, if serious, can be reduced by the use of backflush on the gas chromatograph.
Further, the solvent should not contain impurities which elute around the times of
elution of the peaks which are to be measured.

The desorption of substances by solvent is not always complete, and the efficiency
of desorption may vary from one batch of adsorbent to another, and from laboratory
to laboratory, probably because of slight differences in procedure.   The U.S.
National Institute for Occupational Safety & Health (NIOSH), who have made extensive
studies of solvent desorption procedures, found that, under standard conditions,
the percentages of organic compound recovered from two batches of coconut shell
charcoal of 20/40 mesh by carbon disulphide were as shown in Table 6.2.

It has been shown that the efficiency of desorption of a number of organic substances from charcoal by carbon disulphide increases with time of extraction.
(Ref. 8).   Later work on the desorption of vinyl chloride from charcoal suggests
that the time required for optimum recovery depends on the quantity of the substance
which is adsorbed on the carbon, and further, that the temperature and volume of
the solvent used have little effect on the precision of the procedure. (Ref. 11).
During analysis the results must be corrected to allow for the efficiency of

desorption.

TABLE 6.2   Recovery of Organics from Charcoal by
            Carbon Disulphide

	Charcoal A	Charcoal B
	%	%
Toluene	98	100
Methylethylketone	70	91
2-Ethoxyethylacetate	72	90
1-Butanol	46	60
Vinyl chloride	89	71

Mixing charcoal with a solvent such as carbon disulphide results in an exothermic reaction, and there is always the possibility that the heating of the charcoal and solvent may result in the loss of either the solvent or the adsorbed substances, or both.   To avoid errors from this source certain workers cool either the charcoal or the solvent, or both, when they are recovering volatile substances from charcoal. (Ref. 16).   Any loss from this cause can be minimised by adding the charcoal to the solvent and not vice versa.

The advantages of the solvent leaching method of desorption are that it uses only standard laboratory equipment and procedures, and that it produces the recovered substance in a solution which is easily handled.   It is important, however, that the technique is standardised and the procedures adhered to in detail to ensure precise results.

Disadvantages of the method are that solvents of high purity are required, which may in themselves provide toxic or flammable hazards, and the large dilution of the recovered substances in the solvent, which reduces the sensitivity of measurement. Further, the adsorbent cannot be re-used without laborious regeneration.

Solvent desorption techniques are widely used and have been adopted as standard methods by NIOSH, wherever possible, in their Manual of Analytical Methods. (Ref. 17).   The NIOSH technique, which is essentially the same as that used by most other workers, is described later in this chapter.

Desorption by thermal displacement.   For thermal desorption the trapping tube which contains the adsorbent is heated to a temperature at which the adsorbed substances desorb into the vapour spaces within the tube.   Then a stream of gas is flowed along the tube to sweep out the substances for analysis.   The tube is usually heated electrically, either in a small furnace (Refs. 18, 19), or by direct ohmic heating of a metal tube through which a large electrical current is passed. (Ref. 20).

Unless the adsorbent is fully saturated it is likely that there will be a higher concentration of adsorbed substances near the entry, than at the exit end of the tube.   Therefore, the purge gas should flow in the reverse direction from that in

which the sample flowed.   The pattern of flow of adsorbed substances within the
tube is similar to that of a substance in a gas chromatograph column.   The profile
of the concentration of the substance leaving the tube depends on the total amount
adsorbed, and on its distribution within the adsorbent.   Typical profiles are
shown in Fig. 6.3.

The gas flow is started at time T_0 after an initial preheating time.   Curve A
shows the concentration of the substance in the gas stream leaving a tube in which
a small amount of the substance has been trapped near the sample inlet.   Curve B
shows the profile of concentration of the substance leaving a tube which has contained a larger amount, some of which has been trapped far along the tube.   Curve
C shows the effect of desorbing a tube by flowing the purge gas in the wrong direction, that is, in the direction of the flow during sampling.

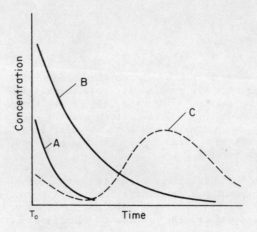

Fig. 6.3 Profiles of concentration of substances leaving a
         trap after thermal desorption. T_0 = time of start
         of gas flow; for explanation of curves A, B and C
         see text.

In flash desorption with a gas chromatograph analyser, the effluent from the
desorber is injected directly into the gas chromatograph.   It is, therefore, important to ensure that the substance leaves the trap within a time which is as
short as possible.   If not, the peak recorded by the chromatograph will be seriously distorted and subject to severe tailing, which may lead to error in measurement.
The tailing can be overcome if the eluted substance is trapped as it leaves the
tube and later injected into the chromatograph rapidly.   This can be done by con-
densing the substance in a cold trap which is later heated to vaporise the substance
before injection.   Alternatively the desorbing gas and the desorbed substance can
be retained in a suitable vessel from which aliquots can be taken for subsequent
analysis.   In this system the pattern of elution from the trapping tube is unimportant, and desorption conditions can be chosen to simplify or optimise the
operation of the equipment.

The advantages of thermal desorption include the absence of solvents, the elimin-
ation of handling of the adsorbent, more complete recovery of adsorbed substances
than is possible with solvent desorption especially when sampling polar compounds,
and little or no dilution of the substances on recovery.   Further, the trapping
tubes can be re-used after desorption, and the sample is recovered in the vapour
phase suitable for analysis by gas chromatography, spectroscopic methods or colour
indicator tubes.

The disadvantages of the method are that special desorption equipment is required
and, when simple desorption equipment is used, only one measurement is possible
from each trap.   Therefore, if the analytical equipment is not correctly adjusted
the entire process of sampling and measurement is abortive.   This is not always
as serious a drawback as it first appears because, in practice, in most monitoring
problems the range within which the measurement is likely to fall is known and the
equipment can be set up accordingly.

Thermal desorption techniques have been developed and used by a number of workers.
(Refs. 16, 20, 21, 22).   However, two systems of trapping followed by thermal
desorption, designed and made respectively by Environmental Monitoring Systems
(Ref. 18) and Century Systems (Ref. 19), are commonly used in U.K.   The procedures
used in these systems are described later in this chapter.

Head-space analysis of adsorbed substances.   Head-space analysis is a technique
normally used for the measurement of gases which are in solution in liquids.   The
liquid is sealed into a vial which is closed by a rubber septum.   The vial is
placed in a bath which is thermostatted at a temperature which is selected according
to the analysis being undertaken, and the liquid and vapour phases in the vial are
allowed to come into equilibrium at this temperature.   A sample of the vapour from
the vial is then extracted, either manually using a hypodermic syringe, or automatically using a mechanised sampler.   The vapour is then analysed, usually by gas
chromatography.   The procedure is calibrated using solutions of known concentration,
and from this calibration data the concentration of the Measured component in the
original solution can be calculated.

The advantage of head-space analysis over direct gas chromatography arises from the
fact that the concentration of the gas in the vapour space is greater than that in
the liquid phase.   It is particularly beneficial at very low levels of concentration when the additional stage of concentration gives a greater sensitivity of
measurement and allows a lower limit of detection to be achieved in the analysis.

The technique can be extended to the analysis of gases adsorbed on solids, and has
been used to determine the vapours adsorbed on charcoal during atmospheric monitoring.   However, calibration involves the determination of the adsorption equilibrium
of the vapours on the charcoal and this can be a difficult and lengthy process which
is justified only if large numbers of samples have to be analysed.   The calibration
problems are considerably simplified if the desorption is done by solvent.   In this
case the calibration involves measurement of the partition equilibrium between the
gas and the solvent, which is easily done by analysis of solutions of known concentration.   The analytical procedure then is to place the charcoal in the vial, add
a measured volume of solvent, seal the vial and equilibrate at the appropriate
temperature.   The charcoal has no significant effect on the equilibrium between
the liquid and gas phases, and so analysis of the vapour in the head-space can be
used to calculate the amount of the measured component which is present in the
solvent.   If desorption is complete, this is the amount of substance which was
present on the charcoal. (Ref. 23).

The NIOSH System of Analysis by Adsorption

The U.S. National Institute for Occupational Safety and Health (NIOSH) has over a
number of years developed methods of environmental analysis many of which are based
on the adsorption of pollutants in special traps, followed by their desorption into
a solvent for measurement.

The NIOSH methods are published in a Manual of Analytical Methods. (Ref. 17).   The
following description of the NIOSH techniques of selective adsorption and desorption
is based on the methods which appear in that manual.

Trapping system.   Where possible, NIOSH use charcoal for collecting the sample.
For this purpose they have designed a special trap which is available commercially
in a number of sizes. (Ref. 24).   The standard trap (Fig. 6.4) consists of a glass
tube, 4 mm internal diameter and 70 mm in length.   It contains two unequal beds
of charcoal separated from each other by a plug of polyurethane.   A polyurethane
plug retains the smaller bed, but the plug at the other end, which is the inlet, is
of silylated glass wool.   The larger adsorbent bed contains 100 mg, and the smaller
50 mg, of coconut shell charcoal of 20/40 mesh which has been activated by baking
at 600°C immediately before being inserted into the tube.   The ends of the tube
are sealed after it is filled.

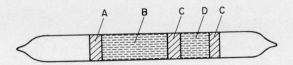

Fig. 6.4   Charcoal sample collection tube, NIOSH design.
          A = silylated glass wool plug;   B = 100 mg coconut
          shell charcoal;   C,C = polyurethane plugs;
          D = 50 mg coconut shell charcoal.

In use, the ends are broken off the tube to give an opening at least 2 mm in
diameter and the sample is drawn through the tube at an appropriate flow rate.
The sample passes first through the large bed of charcoal and then through the
smaller bed which is used as an indicator of breakthrough.   If more than 25% of
the total substance recovered is trapped on the second bed of charcoal it is possible that loss of sample has occurred.   The pressure drop across the tube is less
than 1 inch of mercury at a flow rate of 1 litre per minute, or 2 inches of water
at 200 ml per minute.

The minimum volume of sample which must be flowed through the tube is that which
will provide sufficient material for a measurement to be possible at the lowest
atmospheric concentration which it is desired to measure.   For organic solvents
and similar substances NIOSH have applied the standard that measurement should be
possible down to one tenth of the concentration specified by the U.S. Occupational
Safety & Health Administration (OSHA) as the maximum concentration permitted for
the substance in industrial applications.   The maximum volume of sample which can

be taken is limited by breakthrough from the charcoal bed.   This volume depends on
the concentration in the atmosphere of the substance which is being trapped, and
the capacity of the analysis bed for the substance under the conditions of sampling.
For organic solvents NIOSH have adopted the standard that the system should be able
to measure concentrations up to five times the OSHA standard.   On this basis the
sample volumes for collecting selected organic solvents are as shown in Table 6.3.
An allowance should be made for the effects of high humidity by assuming that the
capacity of the bed is reduced to one half.   From these values the rate of flow
of the sample to give the desired sampling time can be calculated.   This, however,
should not exceed 1 litre per minute.   The flow rate and sampling time, or the
total volume sampled, must be measured, and the minimum volume specified must be
sampled to allow the specified sensitivity to be achieved.

TABLE 6.3   Sampling Volumes for Organic Solvents using
NIOSH Charcoal Tube

Recommended Sample Volume (litres)

	Minimum	Maximum
Acetone	0·5	7·7
Benzene	0·5	55
Carbon tetrachloride	10	60
Chloroform	0·5	13
Dichloromethane	0·5	3·8
Ethylene dichloride	1	12
Methyl ethyl ketone	0·5	13
Tetrachloroethylene	1	25
1,1,2 - trichloroethane	10	97
Toluene	0·5	22

For the collection of vapours which are not efficiently collected by the standard
tube special procedures are adopted.   For example, for collecting vinyl chloride
NIOSH use two standard tubes in series.   For collecting methyl chloride a tube
similar in design to the standard tube but much larger in size is used.   This tube
contains 2 g of charcoal in the analysis bed and 0·5 g in the second bed.

For certain substances NIOSH has found that coconut shell charcoal is too highly
activated and that difficulty arises in the subsequent analysis.   Therefore, for
example, for collecting ethylene chlorohydrin and 1.1.2.2 tetrachloroethane they
specify traps of standard design which are filled with the less active petroleum
based charcoal.

Certain substances cannot be trapped and analysed effectively using charcoal as the
adsorbent, and for them NIOSH specified other adsorbents.   For example, while
acrylonitrile can be trapped by the standard charcoal tube provided the concentration is within a fairly narrow range, NIOSH have investigated a method of trapping

acrylonitrile on Carbosieve B (a type of pyrolysed Saran) which covers a much wider range of concentrations.   For the collection of aromatic amines they specify a tube of unusual design which contains silica gel as adsorbent.   It contains three beds of silica gel adsorbent, two of 150 mg and, at one end, a bed of 700 mg.   The sample is flowed through the tube in a direction which depends on the concentration of aromatic amines, the humidity of the sample and duration of sampling.   Samples of high concentration, high humidity or long duration are passed through the tube in the direction in which they flow first through the large bed of adsorbent. Samples with low concentrations, low humidity and of short duration are flowed in the opposite direction.   By this means the sample is trapped in a bed of appropriate size.   Tubular glass separators between the beds significantly reduce the migration of the sample throughout the tube after sampling and before analysis.   The recom- mended sample volumes for sampling with this tube to monitor the aromatic amines for which the method has been tested are shown in Table 6.4.

TABLE 6.4   Sampling Volumes for Aromatic Amines using
NIOSH Tube

	Recommended Sample Volume (litres)	
	Minimum	Maximum
Aniline	5	15
N,N - Dimethylaniline	5	190
o - Toluidine	5	300
2,4 - Xylidine	5	430
o - Anisidine	250	35,800
p - Anisidine	250	33,000
p - Nitroaniline	80	12,100

Other adsorbents are proposed or used by NIOSH for special purposes.   Examples of these appear in Table 6.5.

TABLE 6.5   Adsorbents other than Charcoal and Eluent
used by NIOSH

Substance	Adsorbent	Eluent
Acrylonitrile	Carbosieve B	Methanol
Aromatic amines	Silica gel	Ethanol
Formaldehyde	Alumina	1% methanol in water
Bis (chloromethyl) ether	Chromosorb 101	-
Sulphur dioxide	Molecular Sieve	-
Nitroglycerine	Tenax GC	Ethanol

For sampling, the tubes are opened by breaking off the sealed ends to provide an opening of at least half the internal diameter of the tube.   The sample must flow in the correct direction and the tubes should be used in a vertical position to avoid channelling.   The sample should not pass through any tube or hose before entering the tube.   The sample volume, and atmospheric temperature and pressure must be recorded.   After sampling is completed the tubes must be capped with caps of plastics material (not rubber).

Desorption system.   In most of the methods developed by NIOSH the adsorbed substances are recovered from the tube by a solvent appropriate to the adsorbent and substance being measured.   However, thermal desorption is used in certain special cases.   For example, a method proposed for the measurement of sulphur dioxide in air involves trapping the sulphur dioxide on a Molecular Sieve from which it is recovered by thermal desorption and measured by mass spectrometry.

For desorption by solvent the beds of charcoal tubes are removed individually from the trapping tube, and each bed is placed in a separate stoppered test tube.   A measured volume of the chosen solvent is added to each tube, and desorption is complete in about 30 minutes if the tubes are shaken gently from time to time.   A variation of this procedure is specified in the determination of vinyl chloride. In this case, the carbon with the adsorbed vinyl chloride is added to the measured volume of solvent and not vice versa.

The solvent favoured by NIOSH for desorption from charcoal is carbon disulphide. This is available in a pure form, and gives only a small response in the flame ionisation detector of the gas chromatograph which is the method of analysis used where possible.   The carbon disulphide must, however, be handled under a fume hood because of its high toxicity.   Hydrocarbons, simple ketones and esters are efficiently desorbed from charcoal by carbon disulphide.   It is not, however, sufficiently active to desorb all compounds from charcoal and, for these, other eluents are required.   For example, carbon disulphide to which a few percent of an alcohol has been added will efficiently desorb alcohols, methanol is used as eluent for methyl chloride, ethyl acetate as eluent for di-ethylether, and tetrahydrofuran as eluent for glycidol.

For desorbing substances from adsorbents other than charcoal a range of eluents is used and examples of these are shown in Table 6.5.

It is stressed by NIOSH that the extraction by solvent is not always complete and the efficiency of desorption must be determined.   This is done by loading known amounts of the substance which is being measured on aliquots of the charcoal, and processing the charcoal using the procedure employed during the desorption of the samples.   The efficiency of desorption can then be calculated and this factor is used in computing the results of analysis.

In some cases NIOSH specifies the use of an internal standard in the solvent used for desorption.   When this is not used a special procedure of solvent flush injection is used in the transfer of aliquots from the desorption vial to the gas chromatograph.   In this technique a 10 µl injection syringe is flushed several times with solvent.   Three µl of solvent are then drawn into the syringe to increase the accuracy and reproducibility of the injected volume.   The needle is removed from the solvent and about 0·2 µl of air is drawn in to separate the solvent from the sample.   The needle is inserted in the sample and a 5 µl aliquot is drawn into the syringe, taking into consideration the volume of the needle since the entire sample will be injected.   The needle is removed from the sample and a small amount of air is drawn into the needle to minimise evaporation from the tip. Duplicate injections are made and the peak areas for the two injections should not differ by more than 3%.

The Environmental Monitoring Systems Equipment

This system was designed by and is manufactured by Environmental Monitoring Systems. (Ref. 18).

The system is based on three standard types of vapour trapping tubes together with a thermal desorption unit which is attached to a standard gas chromatograph. The three tubes are of stainless steel:

$\frac{1}{4}$ inch diameter, containing 550 mg coconut shell charcoal, 40/50 mesh

$\frac{1}{8}$ inch diameter, containing 70 mg coconut shell charcoal, 40/50 mesh

$\frac{1}{4}$ inch diameter containing 300 mg Poropak Q, 100/120 mesh

The adsorption of a number of substances on these three tubes has been studied experimentally. It has been shown that up to certain values the breakthrough volume is independent of concentration, and that the logarithm of the breakthrough volume is linearly dependent on the reciprocal of the absolute temperature for the charcoal adsorbent, as is predicted by adsorption theory. Some deviation from linearity has been shown with the Poropak adsorbent which indicates the occurrence of chemisorption.

The data determined for a number of normal alkanes were analysed to give a single mathematical expression which describes the breakthrough volume in terms of boiling points of the normal alkanes. From this the breakthrough volume for other normal alkanes can be predicted. It was found that the same relation predicts reasonably well the breakthrough volumes of other hydrocarbons, but substances which contain other functional groups have breakthrough volumes which deviate from the calculated value. This is to be expected from theory, from which it can be shown that the heat of adsorption should be used for the correlation, rather than the boiling point. However, the value of the heat of adsorption for many substances is not readily available and, of the thermodynamic functions tested, the boiling point gave the best compromise.

However, the correlation can be improved if a quantity called "the effective boiling point" is used instead of the true boiling point. This effective boiling point is derived from the chromatographic retention time of the substance on a column of the adsorbent, relative to retention times for the normal alkanes on the same column. It is determined by replacing the column in a gas chromatograph by the appropriate vapour trapping tube and adjusting the gas chromatography conditions to give a retention time for the substance of about 5 minutes. The retention times for normal alkanes under the same conditions are measured and plotted against their actual boiling points. The effective boiling point for the substance is then read from the graph as the boiling point indicated by the retention time observed for the substance.

Examples of effective boiling points for a number of substances are shown in Table 6.6.

The sample which has been collected in the vapour trapping tube has to be desorbed later for analysis. In the EMS equipment (Fig. 6.5) this is done by heating the tube to a predetermined temperature, and then passing inert carrier gas through the tube. For reasons of analysis, the volume of carrier gas used should be as small

TABLE 6.6   True and Effective Boiling Points of Organics

Substance	Actual BP (°C)	Effective BP (°C)	
		Charcoal	Poropak Q
Methyl chloride	-24	-6	
Vinyl chloride	-13	-26	
1,3 - Butadiene	-4	4	
2 - Butanone	80	32	61
Benzene	80	53	79
Hexene - 1	63	68	68
Butyl mercaptan	98		99
Isopropylacetate	88		80
Trichloroethylene	86		86
Ethanol	79		13
Formic acid	101		53

as possible, but the volume of gas required to remove completely the adsorbed substances - the desorption volume - is dependent on factors which are similar to those which determine breakthrough volume.   It varies, however, according to whether there is a small amount of substance adsorbed at one end of the bed, or whether there is a large amount adsorbed along the whole length of the bed.   The latter condition requires a much greater volume of gas for total desorption.   The desorption volume for each type of vapour trapping tube has been determined as a function of the effective boiling point of the substance which is being desorbed.

The choice of tube for any particular application is determined by the best compromise between a breakthrough volume sufficiently high to allow sampling to be undertaken at the flow rate selected and for the time required, and a desorption volume sufficiently low to give complete sample recovery and good analytical performance. Weakly adsorbed substances require a tube with a large bed of strong adsorbent ($\frac{1}{4}$ inch charcoal tube) whereas high boiling, strongly adsorbed substances need a less powerful adsorbent ($\frac{1}{4}$ inch Poropak).   However, the useful ranges of these two tubes do not overlap, and to cover the range of intermediate values the $\frac{1}{8}$ inch charcoal tube is used.

Other tubes can be made for special purposes.   For example, Poropak Q does not retain efficiently highly oxygenated compounds, and other types of Poropak must be used.   For substances with higher boiling points a weaker adsorbent such as Tenax GC may be used.

The usable ranges of the standard tubes are shown in Table 6.7.   These values may vary slightly for different batches of adsorbent.

New tubes, and tubes which have not been used for some time, should be cleared of adsorbed substances by flushing them with carrier gas at desorption temperature for 10 minutes.   Tubes can be used repeatedly after desorption.

The sample is collected using any convenient type of sampling pump which will provide the appropriate rate of flow through the vapour trapping tube.   The breakthrough volumes quoted in Table 6.7 are for ambient temperature of 20°C but smaller
sample volumes must be used if the temperature exceeds 20°C.   The reduction
required can be determined from the fact that the logarithm of breakthrough volume
for any substance is proportional to the reciprocal of the absolute temperature.
The breakthrough volume is independent of the vapour concentration provided the
total volume of vapour adsorbed on the tube does not exceed 3 ml on $\frac{1}{4}$ inch tubes
or 1 ml on $\frac{1}{8}$ inch tubes.   Above these values the breakthrough volume decreases

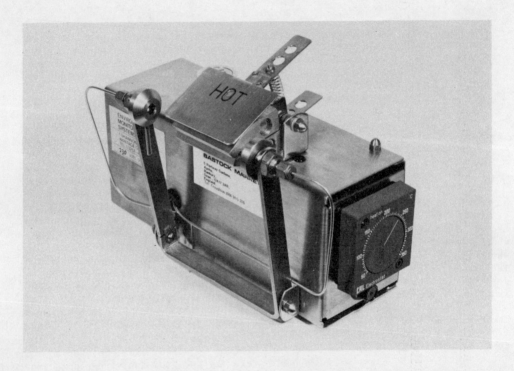

Fig. 6.5 Environmental Monitoring Systems Vapour Desorption Unit.

(Courtesy of Bastock Marketing)

with concentration.   The breakthrough volumes quoted are maximum values and will
be reduced by humidity.   In practice a safety margin for humidity and tube-to-tube
variation should be allowed, and the total sample taken should be limited to about
60% of the quoted values of breakthrough volume.   The sampling flow rate should
be adjusted to take a quantity approximating to, but not exceeding, this volume,
in the time over which sampling is required.   If these precautions are observed
there is no need for a second bed of adsorbent to check for breakthrough.

Desorption is carried out on a special unit which is attached to a gas chromatograph.   This unit enables carrier gas to be diverted through the vapour trapping
tube after it has been heated by a high capacity block heater to a predetermined
temperature.   The desorbed sample can be passed directly into the column of the

TABLE 6.7 Range of use of EMS Vapour Trapping Tubes

Vapour Trapping Tube	Effective Boiling Point (°C)	Compound	Breakthrough Volume (litres)	Desorption Volume (ml)
¼ inch 550 mg charcoal (Effective Boiling Range -90°C to + 20°C)	-88	Ethane	0·3	4 – 6
	-42	Propane	9	6 – 12
	-26	Vinyl chloride	20	8 – 16
	- 1	n – Butane	300	15 – 37
⅛ inch 70 mg charcoal (Effective Boiling Range 0°C – 80°C)	- 1	n – Butane	6	3 – 5
	-26	Acrylonitrile	30	4 – 10
	36	n – Pentane	45	4 – 13
	69	n – Hexane	600	5 – 37
¼ inch Poropak Q (Effective Boiling Range 70°C – 150°C)	36	n – Pentane	1	7 – 15
	69	n – Hexane	3	8 – 19
	80	Benzene	6	9 – 21
	98	n – Heptane	13	12 – 24
	126	n – Octane	45	21 – 40

gas chromatograph.   The vapour trapping tube is loaded into the desorber and
attached by couplings which make a leak-tight seal with the carrier gas system, in
such a way that the gas flows through the tube in the opposite direction to that in
which the sample flowed.   Desorption temperatures of up to 250°C can be used with
Poropak and up to 300°C with charcoal.

If the analytical procedure requires more than one aliquot of sample gas, the
desorption gas may be collected in a gas syringe which can be attached to the
system, and from which aliquots may be taken for injection into the gas chromatograph column.

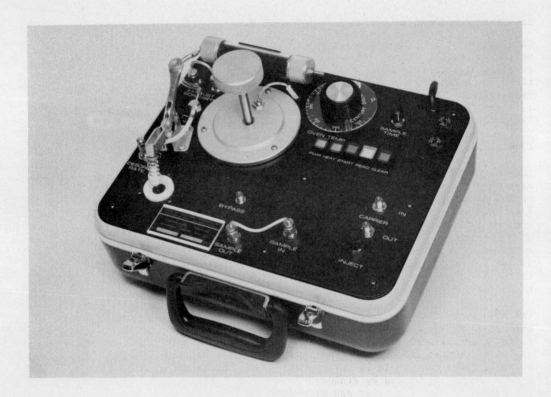

Fig. 6.6 Century Systems thermal desorber.

(Courtesy of D.A. Pitman Ltd)

The desorbed gas may be analysed by techniques other than gas chromatography, for
example by infra-red spectroscopy or by colorimetry.   A special attachment 's
available for desorbing the contents of the vapour trapping tube in such a way that
they can be analysed by a colour indicator tube.   The unit, which is easily portable, is operated by a mains electrical supply and is ready for operation about
five minutes after switching on, when the heater has reached working temperature.
An indicator tube is opened and inserted into a connector, and the vapour trapping
tube which contains the sample is loaded into the desorber.   The flow rate from
an air pump which is incorporated in the unit is set according to the type of tube

in use.   The vapour trapping tube is inserted into the heater and the trapped
vapour is released and flushed by the air into the indicator tube.   The resulting
colour change is a measure of the amount of vapour in the sample.   The reading on
the tube is corrected for the volume of air sampled to give the concentration of
the vapour in the original atmosphere.   The analytical procedure takes less than
one minute and after the analysis the vapour trapping tube is ready for re-use.
There is, of course, no separation stage as is present in gas chromatograph analysers, and so the possibility of errors arising from the presence of interfering
substances must be considered.   Further, the indicator tube is used under conditions
of flow and temperature different from those under which it was calibrated by the
manufacturer.   It should, therefore, be re-calibrated under the conditions used
in the thermal desorber system.

The Century Programmed Thermal Desorber System

This equipment is marketed by Century Systems. (Ref. 19).   The system is an alternative to the EMS system and works on principles which are basically similar.   It
consists of stainless steel sample trapping tubes, 3 inches in length, which are
packed with charcoal or a porous polymer adsorbent, together with a thermal desorption system.   The tubes are known as Flare Vapour Sampler Tubes because one end
is flared to produce a gas-tight seal when the tube is inserted into the desorber.
Tubes can be connected together to form a back-up section to check for breakthrough
if required.   The tubes can be used with any sampling pump which can provide a
flow at the required rate through the tube.

The desorber (Fig. 6.6) consists of a heating system by which the flare tube is
heated to a predetermined temperature.   Air, which has been passed through a
carbon scrubber to remove organic impurities, or, alternatively, an inert gas, is
preheated and then passed through the flare tubes at a rate of about 60 ml per
minute.   The effluent from the flare tube is collected in a stainless steel cylinder of volume 300 ml, which contains a piston which has a seal of polytetrafluoroethylene.   When the cylinder is filled with effluent gas the flow of desorbing gas
is stopped, the temperature of the gas in the cylinder is allowed to return to
ambient, and then aliquots can be taken from the cylinder for analysis.   The cycle
of operations is controlled automatically.

The advantages of this system are that the sample which is injected into the gas
chromatograph has a uniform composition and so the peaks on the chromatogram do not
suffer from tailing.   Again, several aliquots can be taken from the sample if the
analysis requires different chromatographic conditions for different components.
Further, since the sample is uniform and at an ambient temperature the gas chromatograph can be calibrated by standard procedures.   Finally, the system of heating
the tube with heated carrier gas eliminates the possibility of overheating the
contents of the tube with consequent oxidation or degradation.

PASSIVE SAMPLE COLLECTORS

All the selective sample collectors described so far in this chapter require means
for pumping the sample into or through them.   This pumping device complicates the
system, increases its weight, and demands a source of power.   These factors are
very important when the system has to be portable, and especially in personal
monitoring when the sampling equipment must be carried by a worker while he carries
out his normal duties.   A number of selective sample collection systems have been
described which do not require a sampling pump and have been called passive samplers.
These are, in principle, very simple and are attractive for personal monitoring
applications, and also for monitoring at remote sites where power is not available.

Development of these systems is still proceeding.

Diffusion Sampler

This device uses the principle of molecular gas diffusion into a chamber of known geometry.   It was described as a personal monitor for sulphur dioxide (Ref. 25) and later developed as a monitor for nitrogen dioxide. (Ref. 26).   The diffusion chamber is a flat bottomed cylinder on the base of which is a reagent or solvent which will trap the required substance.   Under these conditions, the concentration of the pollutant in the air at the bottom of the tube can be assumed to be zero. The rate of diffusion of the molecules of pollutant into the sampler is then determined only by the dimensions of the diffusion chamber.

The equation for unidirectional diffusion of a gas G1 through a gas G2 at constant temperature is

$$J = D \ dx/dz \qquad\qquad\qquad (1)$$

where J = diffusion flux in moles/cm^2/s

D = diffusion coefficient of gas G1 through gas G2

in cm^2/s

x = concentration of diffusing gas G1 in moles/cm^3

z = distance in direction of diffusion.

If the concentration of gas G1 at one end of a tube is maintained at zero value and the other end is exposed to the ambient air the following relations apply at the ends of the tube:-

where $z = 1$, $x = 0$

and where $z = 0$, $x = c$

where 1 = length of the tube

c = concentration of gas G1 in the atmosphere.

Then from equation (1)

$$J = Dc/1 \ moles/cm^2/s \qquad\qquad\qquad (2)$$

If the cross-sectional area of the tube is A cm^2 and Q is the quantity of gas G1 which diffuses in time t seconds, then

$$Q = J \ A \ t \qquad\qquad\qquad (3)$$

$$= DctA/1$$

that is $c = Q1/DAt$ \qquad\qquad\qquad (4)

Of these factors, 1 and A depend on the size of the tube, D depends on the substance which is diffusing, t is the sampling time and Q is the total quantity of substance which is collected in that time.   Therefore, the device can be used as a quantitative sampler.

It is, of course, necessary to know the diffusion coefficient of the gas which is being collected. The method used for estimating the coefficient for nitrogen dioxide in air is described in Reference 26. The diffusion coefficient is

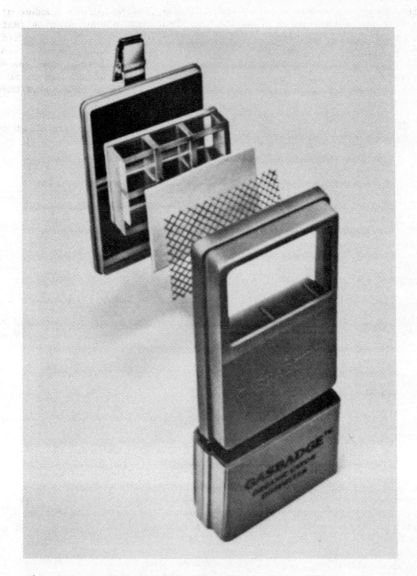

Fig. 6.7 Gas Badge passive sampler.

(Courtesy of D.A. Pitman Ltd.)

proportional to absolute temperature to the power 1.5. However, the volume of a gas is directly proportional to the absolute temperature and so the concentration per unit volume is inversely proportional to this temperature. Therefore, the amount of gas transferred along the tube is proportional to the absolute temperature

to the power 0·5.   Thus the correction required for a change in temperature from
that at which the coefficient of diffusion is calculated is small.

It is important that the concentration at the closed end of the diffusion tube
should be as near to zero as possible.   This requires an effective trapping
arrangement, and an irreversible chemical trap is provided in the sulphur dioxide
and nitrogen dioxide samplers.   Further, the concentration at the open end of the
tube must remain that of the ambient atmosphere.   Under completely still conditions
the diffusion process tends to remove the gas from this point, and.so some movement
of the air over the tube is required.   The dimensions of the tube must be chosen so
that the diameter is large enough not to inhibit the diffusion process, and the
length great enough for the diffusion to be the predominant means of transport of
gas along the tube.

A sampler designed on this principle is commercially available (Ref. 27) and is
described in US patent 3985017 of 1976.   In this design a multi-tube matrix covers
the trapping layer.   The other end of the tubes are covered by a thin permeable
protective membrane.   The tubes are a little over 1 cm in length but it is claimed
that, with the arrangement provided, errors introduced by convective transfer and
air motion are negligible.   The sampler has dimensions of 50 x 60 x 20 mm and
weighs less than 30 g.   The response time is of the order of 1 minute.

In a later development of this device by the same company (Fig. 6.7), the chemical
trap, which had to be specific for each vapour measured, is replaced by a charcoal
collection element.   By this means any substance which can be retained on charcoal
can be collected and the device is suitable for collection of organic vapours.   It
can be used for the collection of samples over 8 hours, and the suppliers claim an
accuracy of about ± 25% in the measurement of time-weighted average concentration.
Where the sampler is used the air velocity must exceed 0·08 metres per second, which
is less than the velocity of air in chemical plants and laboratories under normal
working conditions.

The vapour which is trapped on the charcoal is desorbed by solvent, using a modification of the NIOSH technique, and measured by any suitable analysis system,
usually gas chromatography.

A new diffusion sampler has recently become available. (Ref. 28).   This consists
of an open plastics capsule 40 mm in diameter and 10 mm in depth. (Fig. 6.8).   The
collection element is a disc of charcoal foil which lies on the bottom of the
capsule.   The diffusion space above it is defined by a plastics spacer and is
closed by a membrane of porous polypropylene which is retained in position by a
plastics ring.   The sampler is kept in a sealed foil-lined package until required
for monitoring.   After the sampler has been exposed the membrane and its retaining
ring are removed and replaced by a sealing cap which has two small holes which are
closed by plugs.   With this closure cap in place the sampler is sealed and trapped
gases cannot escape.   Analysis of the adsorbed gases is by solvent desorption and
gas chromatography.   The desorption is carried out within the capsule by removing
one of the plugs in the closure cap and injecting the required volume of desorbing
solvent into the capsule.   The plug is replaced and the sample is desorbed for 30
minutes.   Then the two plugs are removed and the solution is transferred to a vial
for analysis.   This technique minimizes the risk of loss of sample during manipulation.   To avoid errors due to memory effects the samplers are not re-used but
are discarded completely after the analysis.

Permeation Sampler

The permeation passive sampler consists of an open chamber in which is a trap which

will retain the substance which is being collected.   The opening in the chamber
is covered by a permeable membrane.

Fig. 6.8 Organic Vapour Monitor.

(Courtesy of 3M United Kingdom Ltd.)

The flow of vapour through a membrane is described by the equation:

$$F = D A (P_1 - P_2)/S \qquad\qquad (5)$$

where   F = the permeation flux of the vapour, through the membrane

             D = a constant at a given temperature which depends on the
                 diffusion constant of the vapour

A = the area of the membrane

S = the thickness of the membrane

P_1 and P_2 = the partial vapour pressures of the vapour on
    the two sides of the membrane.

Provided the trap is efficient the pressure within the chamber of the vapour which
is permeating the membrane can be assumed to be zero.   If the outside of the
membrane is exposed to the ambient air, the pressure of the vapour on that side of
the membrane is that due to the concentration of the vapour in the atmosphere.
Under these circumstances the mass of vapour which diffuses through the membrane
and is trapped is:

$$Q = ct/K \qquad\qquad\qquad (6)$$

where   Q = the amount of vapour collected

             c = the concentration of the vapour in the atmosphere

             t = the time of sampling

             K = a constant which depends on the nature of the vapour
                 and the dimensions of the membrane.

A sampler based on the principle has been described (Ref. 29).   The device was
developed for monitoring sulphur dioxide, and a trap which incorporated a chemical
reagent was used.   The constant K was determined experimentally by exposing
samplers to streams of air which contained known amounts of sulphur dioxide. Films
of different materials were evaluated for use as membranes.   Silicone rubber was
chosen because of its high permeability (350 times that of the polyethylene standard), the small effect of temperature on the permeability (decrease of 5% for 10°C
increase in temperature), and the negligible effect of humidity on the permeation
rate.   The time between exposure of one side of the film and the emission from the
other side was short and the emitted gas reached a concentration of 90% of that of
the test gas within 10 minutes.

A similar membrane permeation device is included in the portable monitor mentioned
in chapter 2. (Ref. 30).   In this device the trap is again chemical, but consists
of an electrolyte in which electrochemical reaction generates a current which gives
immediate indication of the mass of substance which is being collected.

As with the diffusion sampler, the range of substances which can be measured by
the device can be extended by providing a general rather than a specific trap.   A
sampler for vinyl chloride has been described (Ref. 31) in which a charcoal trap
covered by a silicone rubber membrane is used.   The device has dimensions of
40 x 50 x 10 mm and is very light in weight.   The vinyl chloride is desorbed from
the charcoal by solvent or thermal displacement and is measured by gas chromatography.   Limits of detection of 50 and 1 ppb have been claimed using the two methods
of desorption.

Another design of sampler uses a disc of charcoal cloth covered by a membrane of
porous polypropylene. (Ref. 32).   Many substances including methylethylketone,
styrene, chlorinated solvents, solvent naphtha, white spirit and halothane have
been sampled using this device.   Analysis was carried out by appropriate technique
after the substance had been removed from the charcoal cloth by solution or extraction with a suitable solvent.

Certain precautions are required in using permeation samplers.   The choice of
membrane is important.   Certain membranes such as porous polypropylene permit
water vapour to diffuse freely through them, whereas others, in which the permeation
process is predominantly by a solution process, are less permeable to water vapour.
The presence of water vapour in the permeation cell may affect the action of the
trap.   Each membrane material must be calibrated experimentally.   The previous
physical and thermal history of the membrane can have considerable effect on its
permeation properties.

As with the diffusion sampler, the permeation sampler must be calibrated and used
in moving air to avoid depletion of the air at the permeation surface.

Permeation is a reversible process and so vapours can pass through the membrane
equally well in either direction.   Therefore, if the vapour is not tightly bound
in the trap it may be lost from the sampler if the device is moved to a situation
at which the ambient concentration of the vapour is very low.   Thus, samplers
should be sealed by a gas-tight cap when sampling is completed, especially if the
substances collected are not strongly bound in the trap.

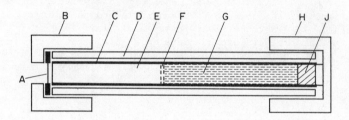

Fig. 6.9 Environmental Monitoring Systems passive sampler.
A = permeable membrane; B = top cap with aperture;
C = vapour absorption tube; D = plastics tube;
E = diffusion space; F = wire gauze; G = charcoal;
H = bottom cap; J = porous plug.

Combined Permeation and Diffusion Sampler

A new passive sampler recently announced (Ref. 33) incorporates both permeation and
diffusion.   The sampler (Fig. 6.9) is based on a modified Environmental Systems
Vapour Adsorption Tube in which is located a disc of wire gauze.   The part of the
tube between the entry end and the gauze forms a diffusion space, the dimensions of
which are critical.   The adsorbent is retained between the gauze and a porous plug
at the other end of the tube.

The tube is contained in a plastics holder when in use, and is attached to the
wearer by a collar clip.   At both ends the holder has screwed end caps, one of
which has an aperture in which is a small disc of a permeable membrane.   The other
end cap can be removed to allow the vapour adsorption tube to be inserted and, when
replaced, seals the end of the tube.   Before and after exposure a special sealing
cap is placed over the membrane to seal the sampler.   After exposure the tube is
desorbed directly on an E.M.S. Thermal Desorber in the normal manner, and the

desorbed vapour is analysed by gas chromatography or other means.   The tube is
ready for re-use after desorption.

No performance data for this device are available at present.

Uses of Passive Sample Collectors

Passive samplers have the attraction that they do not require sampling pumps.
However, their apparent simplicity is misleading and they can present problems in
use.   The design of the sampler must be such that the only air which can reach the
collection element is that which has entered through the limiting device otherwise
it is impossible to calculate the equivalent volume of air from which the trapped
substances have been extracted.   Similarly, when the sampler is not in use, the
collection element must be completely sealed from the atmosphere otherwise errors
may be induced by substances trapped or lost outside the sampling period.   The
calibration of passive samplers presents difficulties for the average laboratory
because these samplers must be calibrated directly by measuring the amount of a
substance which they collect from a moving stream of gas of known composition.

Passive samplers are small, light in weight, and have low initial cost.   It has
been suggested that every worker could wear a sampler when in any area in which he
may be exposed to toxic gases.   However, each time a sampler is used the collection
element must be sent to the laboratory for analysis and this, together with the
necessary calibrations, generates a very heavy analytical load if samplers are used
extensively.   Furthermore, evaluation tests which have been carried out indicate
that the results of concentration measurements made by passive samplers are neither
so precise nor so accurate as those made by pumped sampling and measurement techniques.

Suitably designed passive samplers may, therefore, have applications in making
approximate measurements of concentration, either during an initial survey over a
wide area to find locations of high concentration, or, in a continuing monitoring
programme, to check that conditions have not greatly changed.   They cannot, however, in their present form, replace entirely the technique of active sample collection which must be used if accurate measurements of concentration are required.

REFERENCES

1.   Unbreakable Impinger.   Bastock Marketing, Aynho, Banbury, U.K.

2.   P.W. West & G.C. Gaeke, Fixation of sulphur dioxide as disulphitomercurate
      (II) and subsequent colorimetric estimation, Anal. Chem. 28, 1816 (1956).

3.   R.L. Thomas, V. Dharmarajan, G.L. Lundquist & P.W. West, Measurement of
      sulphuric acid aerosol, sulphur trioxide and total sulphate content of the
      ambient air, Anal. Chem. 48, 639 (1976).

4.   C. Narain, P.J. Manon & J.H. Glover, The analysis of gaseous pollutants,
      Gas Chromatography, 1, (1972).

5.   P. Urone & J.E. Smith, Analysis of chlorinated hydrocarbons with the gas
      chromatograph, Amer. Ind. Hyg. Assoc. J. 22, 36 (1961).

6.   A.P. Altshuller, T.A. Bellar & C.A. Clemons, Concentration of hydrocarbon on
      silica gel prior to gas chromatographic analysis, Amer. Ind. Hyg. Assoc. J. 23,
      164 (1962).

7. W. Thain (Edit), The Determination of Vinyl Chloride, Standard Method SM 10, Chem. Ind. Assoc. Ltd., London, 3rd Ed. (1977).

8. L.D. White, D.G. Taylor, P.A. Mauer, & R.E. Kupel, A convenient optimised method for the analysis of selected solvent vapours in the industrial atmosphere, Amer. Ind. Hyg. Assoc. J. 31, 225 (1970).

9. E.D. Pellizari, J.E. Bunch, R.E. Berkley & J. McRae, Determination of trace hazardous organic vapour pollutants in ambient atmospheres by gas chromatography/mass spectrometry/computer, Anal. Chem. 48, 803 (1976)

10. G.O. Nelson & C.A. Harder, Respirator cartridge efficiency studies, vi. Effect of concentration, Amer. Ind. Hyg. Assoc. J. 37, 205 (1976).

11. R.H. Hill, C.S. McCammon, A.T. Saalwaechter, A.W. Teass & W.J. Woodfin, Gas chromatographic determination of vinyl chloride in air samples collected on charcoal, Anal. Chem. 48, 1395 (1976).

12. A.T. Saalwaechter, C.S. McCammon, A.W. Teass & W.J. Woodfin, Toluene Breakthrough Studies on Activated Charcoal Sampling Tubes, American Industrial Hygiene Conference, Minneapolis, U.S.A., June 1975.

13. W. Thain (Edit), The Determination of Vinyl Chloride, Analytical Note AN 12, Chem. Ind. Assoc. Ltd., London, U.K. 3rd Ed. (1977).

14. Table of Sorbents for Contaminants Listed in ACGIH 1970 Threshold Limit Values, Amer. Ind. Hyg. Assoc. J. 32, 404 (1971).

15. E.J. Otterson & C.V. Guy, A Method of Atmospheric Solvent Vapour Sampling on Activated Charcoal in Connection with Gas Chromatography, Transactions of Twenty Sixth Annual Meeting of the American Conference of Governmental Industrial Hygienists, Philadelphia, Pa. p.37.   A.C.G.I.H. Cincinnati, Ohio, U.S.A. (1964).

16. L.W. Severs & L.K. Skory, Monitoring personnel exposure to vinyl chloride, vinylidene chloride and methyl chloride in an industrial work environment, Amer. Ind. Hyg. Assoc. J. 36, 669 (1975).

17. NIOSH Manual of Analytical Methods, U.S. National Institute for Occupational Safety and Health, Cincinnati, U.S.A. (1977).

18. The E.M.S. System of Vapour Adsorption Tubes and Fast Heat Desorber.   Bastock Marketing, Aynho, Banbury, U.K.

19. Programmed Thermal Desorber.   Century Systems Corporation, Arkansas City, Kansas, U.S.A.

20. D.A. Ferguson & K.J. Saunders, Quantitative Calibration Procedures and Atmospheric Sampling in Environmental Analysis.   International Conference on the Monitoring of Hazardous Gases in the Working Environment, London, U.K., Dec. 1977.

21. S.A. Myers, H.J. Quinn & W.C. Zook, Determination of vinyl chloride monomer at the sub ppm level using a personal monitor, Amer. Ind. Hyg. Assoc. J. 36, 332 (1975).

22. D.H. Ahlstrom, R.J. Kilgour & S.A. Liebman, Trace determination of vinyl chloride by a concentrator/gas chromatography system, Anal. Chem. 47, 1411(1975).

23. B. Kolb, Application of an automated headspace procedure for trace analysis by gas chromatography, J. Chromatography, 122, 553 (1976).

24. Sample tubes for Chemical Hazards in air.   SKC Inc., Eighty Four, Pa. U.S.A.

25. E.D. Palmer & A.F. Gunnison, Personal monitoring device for gaseous contaminants, Amer. Ind. Hyg. Assoc. J. 34, 78 (1973).

26. E.D. Palmer, A.F. Gunnison, J. Dimattio & C. Tomczyk, Personal sampler for nitrogen dioxide, Amer. Ind. Hyg. Assoc. J. 37, 570 (1976).

27. Gasbadge.   Abcor Development Corporation, Wilmington, Mass., U.S.A.

28. Organic Vapour Monitor.   3M United Kingdom Ltd., Bracknell, U.K.

29. K.D. Reisner & P.W. West, Collection and determination of sulphur dioxide incorporating permeation and West-Gaeke procedure, Environmental Sci. & Technol. 7, 525 (1973).

30. Monitox Gas Detector.   Compur Electronic GmbH, Munich, Germany.

31. L.H. Nelms, K.D. Reisner & P.W. West, Personal vinyl chloride monitoring device with permeation technique for sampling, Anal. Chem. 49, 994 (1977)

32. A. Bailey, A Passive Sampler for Gases and Vapours.   International Conference on the Monitoring of Hazardous Gases in the Working Environment, London, U.K. Dec. 1977.

33. The D.D. Personal Sampler.   Bastock Marketing, Aynho, Banbury, U.K.

Chapter 7

SAMPLING PUMPS

<u>INTRODUCTION</u>

Many of the instruments and sampling procedures described in earlier chapters require the atmospheric sample to be made to enter them. In certain samplers, such as the evacuated container and the syringe, the means for drawing the sample into the device is an integral part of its design. In others, some external means must be provided to collect the sample. An atmospheric sample can be drawn into a gas pipette by a simple water-filled aspirator. Such a device, however, is inconvenient to use, operates only for a short time without refilling with water and, because the flow of water leaving the aspirator is dependent on the water level within it, the rate of flow of sample is not constant. Nevertheless, an aspirator can be used to take short term samples if a suitable pump is not available. It has the virtue of requiring no electrical power and so can be used safely in areas where flammable gases may be present.

Generally, samples are drawn into instruments or sampling devices by special pumps which must have certain characteristics according to the particular application in which they are used. They must be able to supply the sample at the flow rate and pressure appropriate to the device to which they are pumping the sample. They must be able to operate reliably for periods of time which may be a few minutes for grab samplers, eight hours or more for time-weighted average samplers, to many days for pumps which are providing samples to fixed monitors. They must be able to operate from whatever power supply is available, which includes batteries for pumps in portable equipment and in devices installed in situations remote from mains electrical supplies. They must not affect the composition of the sample if it flows through the pump before it is collected or analysed. Pumps used in portable equipment must be small and light in weight, and all pumps must be capable of use without creating any hazard in working areas where samples are required.

CHARACTERISTICS OF SAMPLING PUMPS

Sampling Pumps for use with Instruments

If a sample is required by an instrument which makes a direct measurement of concentration of the required component the requirements of the pumping system may be very simple. It must be able to draw or pump the sample into the measurement chamber against such pressure as may be present, and at a rate adequate to ensure that the

chamber is flushed within a reasonable time.   For this application a very simple
pump such as a fan can be used, provided the pressure against which it has to
operate is sufficiently low.

However, most sampling pumps used to provide the flow to an instrument are of the
positive displacement type.   With such pumps the sample flow rate remains substantially constant and independent of small variations in the pressure against which
the pump is operating, provided the motor has an adequate reserve of power.   This
is easily achieved by using motors which are powered by electric supply mains or
substantial batteries.   There is usually no need for the sample flow rate to be
variable, but the pump must be of such a construction that it will not affect the
composition of the sample if it passes through the pump before it enters the
analyser and so the sample must not come in contact with any lubricated part of the
pump.

Oil-free diaphragm pumps are commonly used in this application.   A typical example
is a pump constructed of polypropylene and nylon, and with a diaphragm of nitrile
rubber.   This pump is operated by a 6 volt D.C. motor which has a current requirement of 200 mA, and has a maximum flow rate of 1.2 litres per minute.   Other pumps
of similar design are driven by mains operated motors. (Ref. 1).

If the nature of the sample is such that it cannot be allowed to come in contact
with plastics materials, an all-metal pump can be used.   A pump is available in
which the pumping action is caused by compression and expansion of stainless steel
bellows.   The other parts of the pump with which the gas comes in contact are also
of stainless steel, with the exception of small parts of the valve system which are
of polytetrafluoroethylene.   This type of pump is available in a range of sizes
(Ref. 2).

Personal Sampling Pumps

The demand for personal monitoring by providing a worker with a device which will
monitor continuously, or collect a sample of, the atmosphere to which he is exposed,
has created a demand for small pumps which can be carried by the worker without
inconvenience.   The pump must be reliable in operation, provide a constant flow
rate throughout the sampling period, be small and light in weight, and safe to
operate in areas where flammable gases may be present.   Typical pumps are shown in
Fig. 7.1.

The pumps developed to meet this requirement were designed primarily to collect
atmospheric samples using a selective trapping procedure but they are commonly used
in static samplers which use total sample collection methods.   For example, they
are used to collect samples in gas pipettes.   Strictly speaking, an accurate
metering pump is not required for this purpose because the sample is collected
unchanged in concentration, is taken in a short time and is defined in volume by the
container.   Accuracy of sample flow rate is unimportant, and any simple pump which
can draw the sample through the pipette at a reasonable flow rate will suffice.

When a sample is collected in a bag the volume of sample is not defined by the container and, if the volume is required for subsequent calculation, it is normally
computed from the sampling flow rate and time of sampling.   In selective sampling
procedures the volume sampled must be calculated in this or some similar manner.
In these procedures the sample volume must be known with some accuracy, because this
factor enters into the calculation of time-weighted average concentration of the
component which is being monitored.

If the flow rate of the sample remains constant throughout the sampling period the

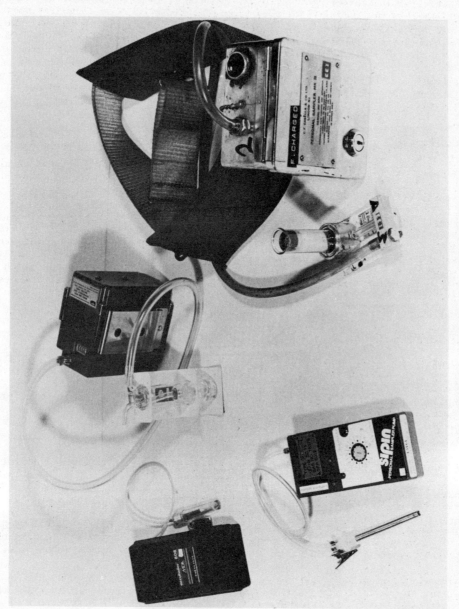

Fig. 7.1 Typical personal sampling pumps with various sampling devices attached

total volume sampled is the flow rate multiplied by the sampling time.   However, there are a number of reasons why the sampling rate may change during the sampling period.   For example, the battery of a portable pump may become partially discharged and the motor speed may fall as a result.   Again, the pressure drop across the sample collection device, for example a charcoal tube, may increase due to an accumulation of particulates.   Further, a sampling tube may become accidently constricted by particulates or by pinching.   These factors may affect the flow rate of a simple pump though, as will be seen, many pumps are designed in such a way as to ensure that pressure changes in the inlet system result in very small changes in the flow rate.   With a well designed pump it can be assumed that the flow rate will remain constant provided precautions are taken to ensure that it is used within the range of the corrective devices.   Nevertheless, it is good practice to check the flow rate of the pump immediately before and after it is used, and to investigate any discrepancy beyond the tolerance quoted by the manufacturer. Certain pumps have flow metering devices built into them which can be used for this purpose and for monitoring the flow during the sampling period.

In positive displacement pumps the total volume pumped is the product of the volume displaced per stroke and the number of strokes made during the sampling period. Certain pumps are fitted with stroke counters for the purpose of calculating total flow by this method.   The accuracy of the result of the calculation depends on whether the volume displaced per stroke remains constant.   This value is dependent on such factors as the efficiency of the valves of the pump, whether they can operate at low pressure difference, and whether they seal efficiently at high pressure difference.   Again, provided that the pump is used within the manufacturer's specification and is in good condition, the total sample volume calculated by this method is reliable.

In certain pumps the metered volume is enlarged and the metering rate is reduced, by metering into some type of bellows.   In this design positive action valves can be used and so the volume displaced by each metering stroke is unlikely to change with external conditions.   This type of pump, therefore, gives an accurate value of total volume sampled, but because of the larger metered volume, the actual flow rate over a short time may not be constant under any conditions of use.

When a pump with a stroke counter is used for sampling, the value of total volume delivered, as calculated from the counter, should be compared with the volume sampled, as calculated from the pump flow rate and time of sampling.   The two values should agree, and any discrepancy exceeding 10% should be investigated.

Some users of selective sampling procedures prefer pumps which accurately record total volume irrespective of flow rate, on the grounds that this is the total volume from which the amount of component measured has been extracted, and so should give the correct value for time-weighted average concentration, even if the flow rate has changed during the sampling period.   This is true only if the change in flow rate is small, or if the instantaneous concentration of the component being measured has not changed significantly during the sampling time.   If there has been a massive increase in the concentration during a time when the sample flow rate was low the method will underestimate the time-weighted average concentration.   The probability of this occurring to an extent which gives rise to serious error is small, but should be considered when time-weighted averages determined by different methods show some divergence.

The flow rate of a personal sampling pump should be adjustable to enable the pump to be used in different sampling techniques.   For example, short term samples may require a flow rate of 200 ml per minute, or even greater, to collect sufficient material for analysis in the sampling time available.   On the other hand time-weighted average measurements over 8 hours with colour indicator tubes require a

flow rate of a few ml per minute.   Most personal sampling pumps provide some means
of adjusting the flow rate either pneumatically, by changing the resistance to the
air flow, or mechanically, by changing the speed of operation of the pump.

In one form of pneumatic adjustment the flow is changed by altering the setting of
a valve, or by changing a limiting orifice, in the inlet of the pump.   If the airflow is restricted in this way the flow rate is reduced but the pressure drop across
the pump is increased.   Adjustment of flow by this method is limited in practice
by the pressure drop which can be tolerated by the pump without the pump losing
efficiency, usually due to leakage at the valves.

An alternative method of pneumatic control is to allow the pump to operate at its
normal pumping speed, but to provide a bleed valve which allows part of the pumped
stream to by-pass the sample collection system.   The flow through the sample
collector can then be reduced by opening the bleed valve.   This system operates
satisfactorily provided the pressure drop across the sample collection device
remains constant after the flow rate has been set.   If the pressure drop changes,
the ratio of the sample stream to the bleed stream will also change.   The extent
of this change can be seen from Figure 7.2.   The test was carried out with a new
pump which had efficient valves which showed no leakage.   The pump had a normal
flow rate of 2 litres per minute and maintained this rate when pumping through a
load across which the pressure drop was adjustable in the range 0 - 5 inches of
water as shown by the top curve.   The pump was then adjusted to pump at a rate of
1 litre per minute by two methods.   In one method the bleed valve was opened to
give a flow rate through the load of 1 litre per minute at zero pressure drop.   As
the pressure drop was increased the volume pumped through the load decreased as
shown by the lowest curve.   The second method of reducing the volume pumped through
the load was to make a mechanical adjustment within the pump to reduce the movement
of the diaphragm.   The bleed valve was closed and the pumped volume again became
independent of pressure drop as shown in the middle curve.   It is recommended that
if consistent flow to high accuracy is required using this type fo flow control,
the pump should be mechanically adjusted to produce a flow rate equal to, or a
little greater than, the value required, and the bleed valve should be used only to
make minor adjustment to flow rate.   This is particularly important if any significant change in pressure drop is likely to take place across the sampler during
sampling.   (Private communication: R. Abrahams, C.F. Casella).

The second method of adjusting flow rate is by altering the working speed of the
pump.   In a few types of pump this is done by altering the stroke of the pump,
usually by changing the stroke of an eccentric or cam.   This involves a mechanical
adjustment to the pump mechanism and cannot readily be done except in an appropriate
workshop.   In most pumps the working speed is adjusted by changing the speed of
the motor which drives the pump.   The manner in which this is done depends on the
type of power supply provided within the sampling pump.

The requirement that a personal sampling pump should be small and light in weight
restricts the size of motor and battery which can be used to operate the pump unit.
It is not possible, as in pumps incorporated in instruments, to use a motor which
has excess power to deal with unforeseen changes in sampling conditions.   Neither
is it possible to use a battery of substantial size and power capacity to provide
high current in the power supply, and to avoid difficulty arising from voltage drop
as the battery discharges.   In a personal sampler the smallest motor which will
drive the pump under the most exacting conditions in the specification must be used,
and the battery must be the lightest which will provide power to the motor for the
maximum sampling time for which the pump will be used.   Miniature electric motors
are used, with small gearboxes to reduce the speed of the drive to a value suitable
for driving pump mechanisms.   These are low voltage D.C. motors and, provided the
load imposed upon them is limited, the rate of rotation is, within quite a wide

range, linearly dependent on the voltage of the power supply.   Thus, if a means of
varying the voltage of the supply is provided this can be used as a speed control
for the motor and as a flow control for the pump.   However, the voltage applied to
the motor must remain constant and must not change as the battery becomes discharged.

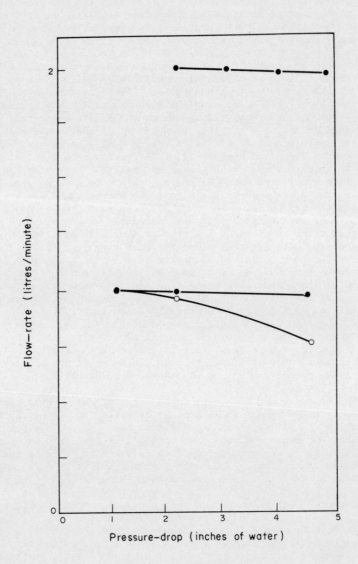

Fig. 7.2   Effect of pressure on flow rate in pump with
by-pass.   For explanation of curves see text.

The power supply for personal sampler pumps is usually a rechargeable battery, but
in certain pumps disposable batteries can be used as an alternative.   When rechargeable cells are fitted the sampler must either be taken out of operation after

use while the cells are recharged, or a fully charged battery must be substituted for the discharged one.   It has been argued that it is simpler to provide a supply of disposable batteries, which are used once and discarded, instead of organising a supply of recharged batteries.

Lead-acid rechargeable cells have been used in certain pumps but nickel-cadmium cells are now more commonly used.   These are lighter in weight than lead-acid batteries of equivalent power capacity, and have certain other advantages which arise from their construction and mode of operation.   In the sealed nickel-cadmium cell the electrodes undergo changes in oxidation state without change in physical state. The electrode materials are insoluble in the electrolyte which is potassium hydroxide.   They do not, therefore, deteriorate significantly even over long periods.   The active material of the positive plate is the hydroxide of nickel oxide (NiOOH) which during discharge goes to a lower valence state $Ni(OH)_2$.   The negative plate contains cadmium metal which during discharge is converted to cadmium hydroxide.   The reactions during charge and discharge are complex but may be expressed:

$$Cd + 2H_2O + 2NiOOH \underset{\text{Charge}}{\overset{\text{Discharge}}{\rightleftarrows}} 2Ni(OH)_2 + Cd(OH)_2$$

No hydrogen is evolved during charging but oxygen can be generated if the battery is overcharged.   However, sealed cells are designed in such a way as to consume this oxygen and so prevent pressure being built up within the cell.   This oxygen reaction absorbs the overcharge current and so prevents damage to the cell.

The nominal voltage of a nickel-cadmium cell is 1·25 volts, and cells can be connected in series to make batteries of higher voltage.   The charging current should be limited by using a voltage of 1·4 to 1·6 volts per cell, to limit the rate of evolution of oxygen should overcharging occur.   The charging current so limited in this way is sufficient to recharge the cell in about 15 hours.   Nickel-cadmium cells may have a memory effect and then do not accept a full charge if they are recharged after having been only partly discharged.   If a battery has been used on intermittent duty it should periodically be completely discharged and then fully recharged.

The characteristics of the nickel-cadmium rechargeable cell have led to its adoption by most manufacturers of personal sampling pumps.   It is light in weight, and has a long operating life.   It can operate for more than 1000 charge-discharge cycles and can be stored for a number of years in a charged condition.   It is tolerant of changing conditions, can be charged from simple charging equipment, and can be kept on a slow overcharge current for long periods without damage.   Large currents can be drawn from the cell without degrading its performance.   The voltage delivered by the cell is nearly constant throughout most of the discharge period.   The cell can be operated in any position and it requires no maintenance in use.

The constant voltage of the nickel-cadmium battery during discharge assists in maintaining the constant motor speed which is required to ensure a constant pumping speed.   However, for high accuracy in flow control, and to obtain the facility for flow adjustment, further control of the motor voltage is necessary.   This can be provided by comparing the motor voltage continuously with a reference voltage derived from another battery, usually a mercury cell.   The difference in these voltages is used to control a regulating circuit which changes the current supplied by the nickel-cadmium battery to the motor in such a way as to maintain the motor voltage at a constant value.   Such a circuit will supply whatever current is required by the motor, up to the practical limit of the batteries, to maintain the voltage required for the desired flow.   The voltage, and consequently the flow rate,

can be changed by adjusting the reference voltage which is used for comparison.
This circuit is suitable for pumping systems in which the changes for which compensation have to be made are not large, and is applicable to sampling pumps used in
monitoring of atmospheric gases and vapours. The requirement for a reference cell
is a complication. In some circuits this can be replaced by a zener diode, operating from the nickel-cadmium cell, with little loss of range of control.

Many applications of personal monitoring require samples to be taken in working
areas where flammable hazards may exist. Most personal monitoring sampling pumps
are, therefore, constructed with limiting circuits so that ignition of gases will
not occur even in the event of a short circuit. The pump case must not, of course,
be opened in working areas. A further safety feature adopted in many designs is to
exhaust the pumped gas into the pump case so that the atmosphere within the pump
is at a slight positive pressure. This is not possible on pumps which have a
pressure outlet for filling sample bags, etc.

Mechanisms used in Sampling Pumps

Various types of mechanisms have been used to provide air displacement in low flow
rate sampling pumps, the most common of which, and their characteristics as pump
elements, are described below. Apart from the fan, all are mechanisms used in
positive displacement pumps.

Fan pump. A small but moderately high speed fan operating in a duct can be used
to pump an atmospheric sample. The device is simple but, because it has no valve
system and relies on the momentum imparted to the air to avoid back flow, it is
limited in application. It can be used to pump against only very small pressure
and the flow is easily stopped by obstruction, even though the fan continues to
operate. It cannot, therefore, be used as a metering pump.

Peristaltic pump. A peristaltic pump operates by compressing a flexible tube
between spaced moving rollers thereby trapping, and moving forward, a sample within
the uncompressed portion of the tube. A pump of this type can be made to operate
at very low flow rates, and the flow rate can be adjusted either by changing the
speed of travel of the rollers or by changing the diameter of the tubing which is
compressed. The device requires no valves because the sealing is performed by the
compression of the tube. The main problems met in using peristaltic pumps concern
the tube. It must be sufficiently flexible to enable the rollers to close it
easily when pumping, but it must return to its original shape rapidly and reproducibly, otherwise the volume pumped will vary. The tube must retain this ability
for long periods and, if any sample is to be drawn through the pump before it is
analysed, the tube must be of a material which will not affect the composition of
the sample. Because of the requirement for complete compression of the tube to
effect a seal the power requirement of the pump is comparatively high.

Piston pump. Piston pumps have been used in high volume air samplers such as are
used for sampling dust and particulates but, because of the problems of construction, are rarely used in low volume sampling pumps. Both single and double acting
pistons have been used, and the mechanism is usually made of plastics materials.
It is usual for either the piston or cylinder to be made of a low friction material,
such as polytetrafluoroethylene, to eliminate the need for lubrication.

Diaphragm pump. This consists of a chamber which is closed on one side by a flexible diaphragm.   A means is provided for moving the diaphragm in such a way as to change the volume within the chamber.   This movement is provided by either an eccentric or a crank driven by a motor, or by an electrical vibrator.   The chamber is equipped with two valves, one to allow air to enter and the other to allow air to leave.   As the diaphragm vibrates air enters and leaves the chamber through the valves.

Diaphragm pumps are reliable in operation and have low demands for power.   A stroke counter is easily fitted to indicate pumped volume.   The flow rate of motor driven pumps can be adjusted, and maintained at a constant value, by controlling the speed of the motor.   This is usually done electrically, though in some pumps mechanical governors have been used.   The flow rate of pumps driven by electro-magnetic vibrators can be varied by changing the stroke of the pump by adjusting the magnitude of the square wave power supply.

The chamber and valve unit are usually made of a plastics material and the diaphragm is usually of polytetrafluoroethylene, neoprene or silicone rubber.   The valves must seal completely and operate on very little pressure difference.

The volume pumped per stroke is usually very small and so any leakage at the valves, or delay in operating due to pressure requirements, can lead to a significant loss in pumped volume.   Because of the critical nature of the operation of the valves they should be protected by a filter from possible fouling by particulate matter.

The flow pattern from a diaphragm pump is not constant but pulses with each stroke. The frequency of pulsation, therefore, depends on the motor speed and the magnitude depends on the movement of the diaphragm.   This pulsation is unlikely to cause difficulty in sampling procedures, but at very low speeds can cause difficulty in reading sensitive flow meters.

Bellows pump.   The bellows pump is related to the diaphragm pump and it is difficult to allocate certain designs to one type or the other.   Again the pumping action is created by changing the volume of a chamber.   In the bellows pump, however, the chamber consists of a bellows or a large diaphragm of hollow shape. The mechanical movement of compression, and the volume of air displaced, are much greater than in the diaphragm pump.   Thus, for the same average flow rate of sample the bellows pump operates at a much lower stroke rate than does the diaphragm pump.

Again the power requirements are comparatively low and, because of the high volume which is pumped at each cycle and the slow stroke rate, the valve requirements are quite different from those of the diaphragm pump.   The flow pattern of a bellows pump is highly dependent on its design and rate of pumping.

Vane pump.   In this type of pump the air flow is created by the change in volume of a section of a cylinder, as a cylindrical rotor rotates eccentrically within it. The air enters and leaves the static cylinder through radial ports, and the seal between the incoming and outgoing air is made by a vane which slides within the rotor or stator,.according to design, as the rotor rotates.   The pump is efficient at high speeds, but there are problems in avoiding air leaks at low speeds.   The vane slides in the slots in the rotor or stator and so must be made of low friction material, such as carbon or polytetrafluoroethylene, to eliminate the need for lubrication.   The flow of gas from a vane pump is almost free from pulsation, and there is no requirement for separate valves because the vane seals the system and prevents reverse flow of gas.

PERSONAL SAMPLING PUMPS COMMERCIALLY AVAILABLE

A wide range of personal sampling pumps is currently manufactured and it is not possible to mention all of them. Those which are described below have been selected to illustrate the various types and their characteristics.

Sipin Personal Sampler Pump (Ref. 3)

This is a diaphragm pump driven by a miniature electric motor and a rechargeable battery. The pump is enclosed in a plastics case with dimensions approximately 30 x 65 x 130 mm, and the weight is approximately 300 g. It is powered by nickel-cadmium cells in a constant voltage circuit which includes a mercury reference cell. The battery will operate the pump for eight hours without recharging. The battery can be recharged within the pump or can be removed easily for recharging which requires about 15 hours. The flow rate of the standard pump can be adjusted by a calibrated control within the constant voltage circuit. The range of adjustment is such that the minimum flow rate is one fifth of the maximum. The flow rate does not change by more than 10% of the initial value set, over the duration of operation of the pump. Provision is made for the connection of an accurate voltmeter into the circuit to measure the motor voltage and allow the flow to be set accurately. Figure 7.3 shows the relation between flow rate and motor voltage for a typical Sipin SP1 pump.

The motor drives a neoprene diaphragm through an eccentric mechanism which also drives a five digit stroke counter. The total volume sampled by the pump is linearly related to the number of pumping strokes within an accuracy of ± 5%. The pump operates at full pumping speed against a pressure of about 20 inches of water. The pump is available with or without an outlet tube.

The valves have been specially designed to operate efficiently with very little pressure drop to ensure calibration linearity over the entire flow range. The sensitivity of the valves makes them susceptible to fouling by dust and this can affect the calibration of the pump. Glass chips from chemical adsorption tubes have occasionally been found in the valves, and so an efficient foam filter is now used in the pump inlet to protect the valves.

Seven models of the standard pump are available each of which has a different range of flow rates. These cover the values of 40 to 200 ml per minute for Model SP1, to 0·5 to 2·5 ml per minute for Model SP7. The pumps are certificated for intrinsic safety by British Approvals Service for Electrical Equipment in Flammable Areas (BASEEFA) and U.S. Bureau of Mines.

The pump is suitable for use with adsorption tubes, time-weighted average colour indicator tubes, and for collecting samples in bags, etc. A modular duty timer can be attached to the pump to extend the period over which a selected volume of air is collected for measurement of time-weighted average concentrations. When the pump is under the control of the duty timer it operates intermittently, being switched on and off for short predetermined times by the timer. The "on" time can be varied in increments of 10% from 10 to 100% of the total time.

A fault alarm can be fitted to indicate an unacceptable drop in flow rate which may arise from a constriction or obstruction in the inlet tube or sampling system. The device senses the suction at the inlet to the pump, and when this rises above a predetermined value the pump is switched off and an audible alarm is sounded. When the fault is rectified the pump can be switched on again immediately.

A recent addition to the range of Sipin pumps is Model SP15 which is generally

similar in appearance and weight to the standard pumps.   The flow rate, however,
covers a much wider range and is continuously variable from 2 to 200 ml per minute.
This is achieved in two overlapping ranges by an electronic speed controller.   It
incorporates a high torque motor which gives a suction of 100 inches of water, and
an overload detector which shuts off the pump and gives an alarm if the inlet load
exceeds a pre-set value.

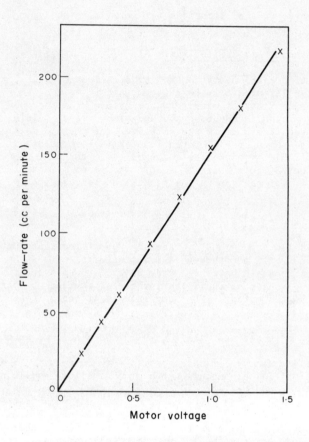

Fig. 7.3   Relation between motor voltage and flow rate on
            Sipin SP1 pump.

(Courtesy of Sipin International Inc.)

Casella I.S. Sampling Pump (Ref. 4)

This also is a diaphragm pump powered by nickel-cadmium rechargeable cells.   It is
contained in a stainless steel case provided with a lock and key to prevent unauthor-
ised opening in hazardous areas.   The overall dimensions are 145 x 115 x 70 mm and
the weight of the complete pump is 1·25 kg.   The battery allows up to 10 hours
operation and is normally recharged in the pump overnight.   The motor, the speed
of which is controlled by a mechanical governor, drives the diaphragm through an

eccentric. A counter also driven by the motor shows the time of operation in
minutes. Flap valves are used on the pump and a pulsation damper is fitted within
the pump case to remove pulsations from the air flow.

The pump was originally designed to pump at a rate of 2 litres per minute for sampling of air for particulate measurement. However, much lower pumping rates can
be achieved by reducing the throw of the eccentric. The desired flow rate is set
accurately by adjusting a bleed valve which allows air to enter the pump stream,
by-passing the sample stream.

The pump is constructed to specifications for intrinsic safety.

Pitman Model 704 Personal Air Sampler (Ref. 5)

This is a double acting piston pump powered by nickel-cadmium batteries. The
dimensions of the sampler are 160 x 80 x 56 mm, and it weighs 540 g. The case is
made of a polycarbonate plastics material, the cylinder of the pump is of polytetrafluoroethylene, the piston, connecting rod and crank are of Delrin, and the valves
are of Terylene. The flow rate is variable over the range 0 to 4·5 litres per
minute and is adjusted by a pre-set potentiometer inside the case. An optional
flow control is available by which the speed of the motor is automatically controlled by a voltage compensating circuit in such a way as to maintain constant pump
flow rate irrespective of changes in the input load on the pump.

Accuhaler Personal Sampler (Ref. 6)

The Accuhaler is a bellows pump and operates by repetitive reproducible inhalations
into a chamber of known volume, through a limiting orifice. The pump is powered
by rechargeable or disposable batteries and, as will be apparent from the design,
the operation is independent of the battery voltage provided the battery has sufficient power to drive the motor. The pump case is of acetal resin and has dimensions of 165 x 90 x 38 mm. The sampler weighs 400 g.

The measurement volume is defined by a flexible rubber bellows which is expanded by
an internal spring. The power is provided by a miniature electric motor which
drives intermittently a cam system which operates two levers. When the bellows
is in the expanded position a contact on top of the bellows completes an electrical
circuit by touching one of the levers. The motor then starts and the following
cycle is initiated. The second lever opens the exhaust valve; the first lever
compresses the bellows against the spring; the second lever closes the exhaust
valve; the first lever springs back to its original position thus opening the
electrical circuit and stopping the motor. The bellows then start to expand under
the influence of the spring, drawing air through the inlet tube which contains a
limiting orifice and filter. When the bellows is fully expanded electrical contact
is again made and the cycle is repeated.

A four digit mechanical counter operated by the cam counts the number of cycles of
operation, each representing one excursion of the bellows. The volume sampled per
stroke is about 4 ml but each pump is individually calibrated. The total count,
therefore, represents the total volume sampled.

The flow rate of the sample is not constant throughout the cycle. It stops while
the bellows are being compressed, starts at a high value and falls as the bellows
expand. The average sampling rate is determined by the limiting orifice which
can be changed as required to give average flow rates in the range of 0·5 to 100 ml
per minute. There is no other means for adjusting the flow rate.

The pump is a total volume meter and the pumped volume reading is independent of flow rate.   Thus if the sampling is interrupted in any way the volume recorded is still valid for the time of operation.   The pump cannot continue to operate in the absence of air flow.   The exhaust valve, the only valve necessary, is power operated and so need not be of high sensitivity.   If the inlet resistance to the pump increases the pump will continue to operate at reduced average flow rate but the total volume sampled will still be accurately recorded.

The pump has been certificated as intrinsically safe in U.K. and U.S.A.

Rotheroe and Mitchell I.S. Gas Sampler (Ref. 7)

This is a small sliding vane pump powered by nickel-cadmium cells.   The dimensions are 120 x 73 x 45 mm and the weight is 425 g.   The battery provides up to 10 hours operation and recharging requires 14 hours.   A red light-emitting diode flashes continuously when recharging is required.   The motor voltage is controlled accurately by an electronic circuit, and the flow is varied by an in-line constriction valve.   A flow meter is provided to set and check the flow rate which can be varied within the range 5 to 500 ml per minute.   At maximum flow rate the pump can operate against a pressure of 7 inches of water.   A filter must be used in the inlet tube to protect the pump against particulate matter.   The sampler has been certified as intrinsically safe by BASEEFA.

Century Portable Air Sampling Pump (Ref. 8)

This pump is of unusual design in that the pumping and metering are carried out by separate units.   A diaphragm pump provides the flow and a bellows system makes the measurement.   An electronic circuit provides pulses at a pre-set frequency which can be adjusted to provide the required flow rate.   Each pulse starts a pumping cycle by starting the diaphragm pump which draws in air and exhausts it into the bellows.   Immediately pumping begins, the vacuum on the input side of the diaphragm pump is used to close a valve in the outlet from the bellows.   The bellows expand against a spring until they reach the fully open position at which they operate a switch which stops the pump.   At this stage the pressure in the bellows is determined by the load of the spring.

When the pump is stopped the vacuum at its input decays, and this allows the exhaust valve of the bellows to open, releasing the air and allowing the bellows to collapse. When they reach the collapsed position they operate another switch, and the pump awaits another timing pulse to start another cycle.   The number of cycles of operation is recorded on a mechanical counter.   Electronic logic circuits are incorporated to prevent malfunctions, such as the commencement of a new operating cycle before the previous cycle is complete.   In the event of malfunction an audible alarm is given.

The average flow rate can be set between 10 and 150 ml per minute and can be maintained against an inlet pressure of 30 inches of water.   The volume of the bellows is 25 ml, and a 4 digit counter indicates the volume pumped from 25 ml to 249 litres. The pump is normally powered by disposable batteries, but rechargeable cells can be used.   Disposable batteries operate the pump for 6 hours at 60 ml per minute against an inlet pressure of 5 inches of water.   Nickel-cadmium cells operate for 10 hours under these conditions.

The pump is cylindrical in shape with a diameter of 60 mm and length of 130 mm. It weighs 450 g.

Draeger Polymeter Pump (Ref. 9)

The Polymeter is a peristaltic pump powered by a lead-acid rechargeable battery which has sufficient capacity for at least 8 hours continuous operation.   The dimensions are 170 x 85 x 50 mm, and it weighs approximately 1000 g.   The constant speed motor drives the pump mechanism through a reduction gearbox which gives a final speed of approximately 20 revolutions per minute.   The pumped volume is approximately 0·75 ml per revolution and a mechanical counter records the number of revolutions made.   From this indication the total volume pumped can be calculated. The flow rate is approximately 1 litre per hour.

The tubing in the peristaltic movement is of special chemical resistant rubber and can be easily replaced when necessary.   The pump is housed in a metal case and has been approved by BASEEFA as intrinsically safe.

The pump was designed for use with Draeger Long Term Tubes for making measurement of average concentrations over times up to 8 hours.   The pump can be carried by the worker on a belt, and an extension hose allows the tube to be fixed within his breathing zone.   It can also be used as a static monitor and in this application the tube is contained within the pump case.

The calibration of the Draeger Long Term Tubes is matched to the flow characteristics of the Polymeter pump.

CALIBRATION OF PUMP FLOW RATE

The flow rate of a pump has to be set to a value appropriate to the application in which it is being used.   The adjustment procedure varies from one type of pump to another, but usually involves either adjustment of a mechanical constriction in the pump inlet, adjustment of a by-pass valve on the airflow, or variation of the speed of operation of the pump.   In certain pumps the flow adjustment device is itself calibrated and is capable of being set reproducibly to give the required value of flow.   Such systems include calibrated potentiometers in electronic flow controllers, and calibrated limiting orifices used to control flow rate.   In other types of pump which do not have calibrated controls the setting must be made with a flow measuring device connected in the pumped stream.   A rapid indication of flow can be made with a rotameter of appropriate range.   However, if the pump, because of its design, gives a pulsing or intermittent flow, it is difficult to interpret the readings from the rotameter.   To achieve high accuracy it is preferable to measure the total sample pumped in a measured time.   If the pumping rate is cyclical or irregular the time over which the pumped volume is measured must be long compared with the pump cycle.

The method of measuring the pumped volume must not impose any additional load on the pump, and a convenient method is to use a soap bubble flowmeter in which the time taken to displace a soap bubble through a known volume is measured.   The known volume is provided by a calibrated glass tube.   A burette is suitable, provided it is of a size which can contain the volume which will be pumped in an easily measured time.   The burette should have no internal constriction which would restrict the gas flow.

The burette is set up in an inverted manner as shown in Figure 7.4.   The beaker under the burette contains a solution of soap or detergent the surface of which is a little way below the end of the burette.   The top of the burette is attached to the pump through a flexible tube in which is a trap which will retain any liquid carried over during operation.   If it is intended to draw samples through any device which will impose a pressure drop load on the pump, for example an absorption

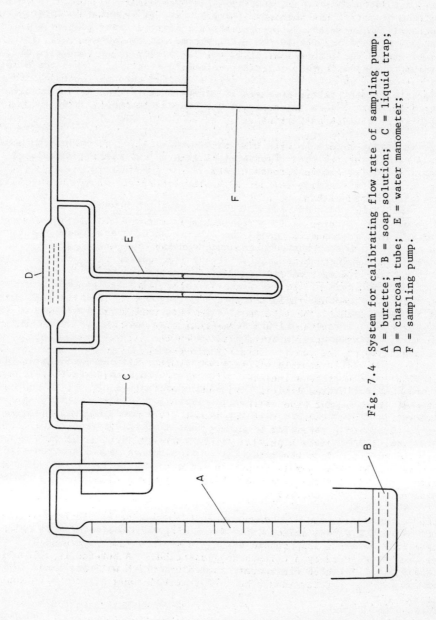

Fig. 7.4 System for calibrating flow rate of sampling pump.
A = burette; B = soap solution; C = liquid trap;
D = charcoal tube; E = water manometer;
F = sampling pump.

sampling tube, a similar device should also be included in the line between the trap and the pump.   If it is required to measure the pressure drop across this device a water manometer can be connected across it.

To use the calibrator the pump is started and a soap bubble is formed in the burette by raising the beaker so that the end of the burette is momentarily immersed in the soap solution.   If the walls of the burette are dry the first bubbles will not travel far up the burette, but as the walls become wet the bubbles will travel over its whole length.   The initial operation may be speeded up if the walls of the burette are moistened with the soap solution before starting.   When the bubbles reach the top of the burette the liquid may be carried through the tube into the trap.   Certain commercially available flowmeters of this type have a bubble breaker which bursts the bubble before it reaches the top of the tube.   With these the trap is not necessary.

The time taken for a bubble to flow between two calibration points of the burette is measured with a stop watch.   The speed of pumping can then be calculated from the volume displaced and the time recorded.

RESPIRATOR COMPENSATION OF PERSONAL SAMPLING PUMP

A personal sampling system will record the time-weighted average concentration of a substance to which a worker has been exposed over the time of sampling.   If, how-ever, he has worn respiratory protection at any time during that period the value measured is not the average concentration of the air which he has breathed.   If the average concentration of the substance in the inhaled air is required, sampling must stop during the time the respirator is worn.   However, respirators are usually put on at a time of emergency and unusual activity , and it is imprudent to require that the worker should switch off his sampler when he puts on the respirator, and to switch it on again when he removes the respirator.

A device for carrying out this operation automatically has been described. (Ref. 10). A microswitch is mounted on the respirator facepiece in such a way that it senses when the respirator is being worn.   This switch controls a circuit in the pump which switches off the pump motor while the respirator is being worn.   It also starts a timer which records the time during which the sampling pump has been stopped.   This time must be known to enable the sampling duration to be calculated. From this value, and the amount of the substance trapped during the time of sampling, can be calculated the time-weighted average concentration of the substance to which the worker was exposed while the respirator was not in use.

REFERENCES

1. Oil Free Diaphragm Pumps for air and gas sampling.   Charles Austen Pumps Ltd., Weybridge, U.K.

2. Stainless Steel Welded Bellows Pumps.   Metal Bellows Company, Sharon, Mass. U.S.A.

3. Personal Sampler Pumps.   Sipin International Inc., New York, U.S.A.

4. Intrinsically Safe Sampling Pump.   C.F. Casella & Co Ltd., London, U.K.

5. Model 704 Personal Air Sampler.   Pitman Instruments, Weybridge, Surrey, U.K.

6. Accuhaler Personal Sampler.   MDA Scientific, Park Ridge, Illinois, U.S.A.

7. Intrinsically Safe Gas Sampler C500.   Rotheroe & Mitchell Ltd., Greenford,
    Middlesex, U.K.

8. Portable Air Sampling Pump.   Century Systems Corporation, Arkansas City,
    Kansas, U.S.A.

9. Polymeter Long Term Pump.   Dragerwerke, Luebeck, Germany.

10. D.E. Moore and T.J. Smith, Respirator compensation of a portable air monitor,
    Amer. Ind. Hyg. Assoc. J. 36, 430 (1975).

Chapter 8

PREPARATION OF STANDARDS FOR CALIBRATION OF INSTRUMENTS AND EVALUATION OF METHODS

<u>INTRODUCTION</u>

Few of the methods of measurement used in atmospheric monitoring are absolute and so they have to be calibrated using mixtures of which the compositions are known. Some devices, for example colour change indicator tubes, are calibrated by the manufacturer, but many analysts check this calibration against their own standard mixtures to ensure that, when used in their method and application, the devices do in fact record correct values of concentration. In addition, the sampling techniques used in environmental monitoring are often complex and tend to be subject to errors, not all of which can readily be foreseen. It is important, therefore, to check the validity of the techniques within the specific applications in which they are to be used. This, again, requires mixtures of known concentrations of the vapours which are to be measured.

Mixtures of gases and vapours at low concentrations in air are available in gas cylinders or disposable containers from laboratory suppliers. Refs. 1, 2, 3, 4, list U.K. Suppliers. Suppliers in other countries can be found in trade directories. Each container is normally supplied to a specification and is accompanied by a certificate of composition of the contents. If the gas or vapour required is supplied in this way, such a mixture is a convenient standard for calibration of equipment, and can also be used for testing methods of sampling. However, if a number of different concentrations are required the purchase of several cylinders is costly and their storage can be inconvenient. A technique for diluting a gas from a cylinder to give mixtures of different concentrations is described later.

When reactive gases such as carbon monoxide or sulphur dioxide are blended with an inert gas and stored in a steel cylinder, the concentration of the gas delivered from the cylinder may vary with time, temperature or pressure. The extent of this variation varies from one cylinder to another and depends on the condition of the internal walls. The main causes of the change are chemical reaction with, or catalysed by, the cylinder wall, and physical adsorption on the wall. To improve the stability of gas mixtures the cylinders in which they are stored must be absolutely clean and dry, free from rust, and as smooth as possible to reduce the surface area. Steel cylinders which have been lined with organic coatings have been used to improve the stability of gas mixtures with some success. Aluminium cylinders have also been used. These are usually rendered inert by the formation of a film of aluminium oxide on the internal walls.

If the gas which is required is not supplied in standard mixtures in this way the analyst must himself prepare mixtures for calibration and testing purposes. Great care must be taken in this operation, and the purest available chemicals should be used, because the accuracy of the results of calibration of the measuring instruments is a major factor in determining the overall accuracy of the results of monitoring. For calibration of instruments only small quantities of the standard mixture are required and, according to the type of instrument, the substance may be required diluted in air or in a liquid. For evaluation and testing of methods of sampling much larger quantities of the standard mixture are required. Methods of preparation of mixtures of gases used by a manufacturer for calibration of colour indicator tubes have been described. (Ref. 5).

Blends of gases can be made by static or dynamic methods. The static methods require simpler laboratory equipment and are suitable for preparing dilute concentrations of many substances in air or nitrogen. However, these methods are subject to errors when they are used to prepare mixtures which contain substances which adsorb readily on the walls of the vessel in which the mixture is contained. The magnitude of the loss of a substance from the mixture by adsorption depends on the surface area to which the gas mixture is exposed and so, in a vessel of given size and geometry, the percentage error in concentration increases as the concentration is reduced. For this reason the method should not be used for the preparation in small vessels of mixtures which contain very low concentrations of substances which are likely to be adsorbed on the walls of the vessel.

Dynamic methods of preparing mixtures of known concentration are much less susceptible to errors arising from adsorption. The areas of surfaces with which the mixture comes in contact can be kept to a low value, and the walls of the equipment can be kept in equilibrium with the mixture which is being prepared. Dynamic methods are capable of producing a mixture of known composition over a considerable period of time and so can be used to test systems designed for long term measurements. Ideally the dynamic system should be capable of being calibrated and checked by some independent and absolute method, preferably by gravimetry, and should be capable of being used to prepare standard mixtures of a wide range of the substances which have to be monitored.

Aliquots of gaseous mixtures which are used for calibration of analytical instruments must be introduced into the instrument by a method appropriate to the method of measurement. The method of transferring the aliquot must not affect the composition of the standard mixture and, for many instruments, must provide a means of accurate measurement of the volume of the aliquot. For introducing a calibration aliquot into a gas chromatograph the gas sampling valve and calibrated loop described in Chapter 5 is convenient.

PREPARATION OF GASEOUS STANDARDS

Static Methods

A mixture containing a low concentration of a gas or vapour of a liquid in air or nitrogen can be prepared by adding a small measured volume of the gas or vapour to a large measured volume of air or nitrogen. The actual volumes used depend on the concentrations required and the apparatus which is available, but the following techniques have given satisfactory results. The procedure should be carried out several times initially to check that reproducible results are obtained.

Flask dilution method. A large glass flask of about 5 litres capacity, and fitted with a ground glass joint on the neck, is used as the mixing vessel and as the large

measured volume.   The stopper of the flask consists of a male ground glass joint
which fits the female joint on the neck of the flask.   The stopper terminates in
a short glass tube which is closed by a rubber septum through which a hypodermic
needle can be inserted into the flask.   Certain workers insert a greaseless stop-
cock between the ground glass joint and the septum.   This can be kept closed to
avoid the contents of the flask coming in contact with the septum, and opened to
allow the passage of the hypodermic needle into the flask.

The volume of the flask is measured accurately by filling it with water and measuring
the water with calibrated flasks and a measuring cylinder.   The flask is then
rinsed with acetone and purged thoroughly with nitrogen or air, according to the
base gas required in the mixture.   If air is used and the supply is likely to contain organic impurities, the air stream should be passed through a large trap filled
with activated charcoal.

The stopper is fitted to the flask.   A measured volume of the gas or vapour which
is to be added to the gas in the flask is taken in a gas-tight syringe by one of
the methods described below.   The syringe is fitted with a long needle which
reaches well into the centre of the flask, and the measured volume of vapour is
injected into the flask through the septum.   A volume of 1 ml of gas or vapour
added to a 5 litre flask will give a mixture which has a concentration of about
200 ppm of the gas.   The contents of the flask must be thoroughly mixed.   This
can be done conveniently using a magnetic stirrer but the stirrer bar within the
flask should be converted to a large paddle by attaching a piece of metal foil.
Stirring should continue rapidly for at least 15 minutes to ensure that the contents
are thoroughly mixed.

Aliquots can be withdrawn from the flask using a 10 ml gas syringe fitted with a
long needle using the following procedure.   The syringe is filled with air and the
needle inserted into the flask through the septum.   The air is discharged into the
flask to pressurise slightly the contents.   A sample is drawn into the syringe and
the plunger is operated several times to condition the walls of the syringe.
Finally the syringe is again filled and the needle is withdrawn from the flask.

The concentration of the gas in the flask is, of course, altered by this procedure
but up to three aliquots can be withdrawn from a 5 litre flask without the concentration being affected to an extent which is significant.

Bag dilution method.   The mixture is made by injecting an appropriate volume of gas
or vapour from a gas syringe into a plastics bag while it is being filled with a
known volume of air or nitrogen.   The bag should be made of an inert plastics
material on which the gas concerned does not readily adsorb.   For most purposes
Tedlar (polyvinylfluoride) is suitable.   The bag should have a capacity of about
20 litres.   About 10 litres of air or nitrogen is flowed into the bag, the volume
being measured accurately by a calibrated flowmeter.   While the bag is being filled
a measured volume of vapour is taken in a gas syringe by one of the methods described below, and injected into the bag through a septum.   If 1 ml of vapour is
injected into 10 litres of air the concentration of the diluted substance in the
mixture is 100 ppm.   The contents of the bag must be thoroughly mixed.   This is
achieved in part by injecting the vapour into the bag while the base gas is still
flowing but must be completed by kneading the bag for about 10 minutes.

Aliquots may be taken from the bag in a gas-tight syringe by inserting the needle
through the septum.   There is no need to inject air from the syringe, as in the
flask method, because the pressure in the bag is not reduced when an aliquot is
withdrawn.   However, the syringe should be operated several times when the needle
has been inserted through the septum, to condition the walls before the sample is

withdrawn.   This procedure does not change the concentration in the bag and so
more aliquots can be withdrawn than can be taken when the flask method is used.

Procedure for filling a gas syringe with gas or vapour.   The techniques described
for preparation of gaseous standards involve the injection of known volumes of
vapour from a gas syringe.   The procedure adopted for filling the syringe depends
on the nature of the substance which is being diluted.   If it is a gas which is
contained under pressure in a cylinder a small quantity of the gas can be transferred to a previously evacuated plastics bag which has a capacity of about 250 ml.
If a septum is attached to the bag a sample can be withdrawn at atmospheric pressure
by a gas-tight syringe fitted with a hypodermic needle.

Instead of a bag, a parallel sided bottle can be used as container for the gas.
This is fitted with a cork through which is a hole or groove large enough to admit
the needle of the syringe, and an inlet tube which reaches to near the bottom of
the bottle.   The gas from the cylinder should be flowed steadily through the bottle
to purge it completely of air before a sample is withdrawn.   If the vapour which
is to be sampled is from a very volatile liquid, the liquid itself may be run into
the bottle and allowed to fill it with vapour by evaporation.   In all methods the
syringe should be operated several times when the needle is in the vapour before a
sample is withdrawn.

The methods which involve removal of gas from a cylinder are satisfactory where the
cylinder contains only gas under pressure.   If the cylinder contains liquified gas,
the vapour which is drawn off will have the same composition as the liquid, provided
the contents of the cylinder do not contain volatile impurities which would concentrate in the vapour phase.   If there is any doubt about the purity of the material
in the cylinder, especially with regard to volatile impurities, the sample should
be taken from the cylinder in the liquid phase and not in the gas phase.

Alternative procedures for preparing mixtures of gases by static methods.   Modifications to the methods described are possible.

A glass mixing vessel which has a capacity of 400 litres is used for the preparation
of gas mixtures for the calibration of indicator tubes. (Ref. 5).   The vessel has
an internal evaporating dish which can be used for the addition of substances in
the liquid phase.   Gases are added by means of gas syringes.   A motor driven
stirrer is incorporated within the vessel and may have to be operated for several
hours to ensure that the contents are thoroughly mixed.   Multi-component mixtures
can be made in this vessel, and mixtures have been made in which some components
have concentrations of less than 1 ppm.

The calibration system provided by the manufacturer of a commercially available
infra-red gas analyser consists of a closed loop which includes the cell in which
the measurement is made, a stainless steel bellows pump, and a septum injection
unit, connected by tubing made of polytetrafluoroethylene.   In operation the pump
creates a continuous air flow around the loop and one microlitre liquid aliquots
of the substance for which calibration is required are injected sequentially
through the septum from a micro-syringe.   The volume of the system is approximately
5.5 litres, and so the addition of 1 microlitre of the liquid gives a mixture in
which the concentration of the vapour is approximately 50 ppm, but the exact value
depends on the molar volume of the gas.   To prepare standards of lower concentration the required substance is diluted in a solvent which will not interfere with
the measurement, and the solution is injected into the system. (Ref. 6).

A gas blender for preparing mixtures of gases of known concentration is commercially

available. (Ref. 7).   The mixing vessel is a glass cylinder of 4 litres capacity
which has a gas connection at each end, and which contains a free piston.   The
piston can be moved along the tube by tilting the tube about a pivot at its mid
point.   Air can be drawn into the cylinder by moving the piston from one end to
the other.   The gas which is to be diluted is measured in a loop of known volume,
and the loop is connected to the ends of the cylinder by the operation of valves.
The piston is then moved repeatedly along the cylinder from one end to the other.
In the course of this operation the gas is swept from the loop into the cylinder
and mixed with the air within it.   An accuracy of ± 1% or 1 ppm is claimed by the
manufacturers for the concentration of mixtures made in this way.

Dynamic Methods

Dynamic methods of producing mixtures of gases or vapours are methods in which the
gases or vapours are allowed to enter a mixing chamber at known rates.   The
effluent from the mixing chamber then contains the various components in concentrations proportional to their rates of flow into the chamber.   When a mixture which
contains a very low concentration of a component is required, special techniques
must be used to produce the very low flow required for that component, and some of
the techniques which can be used are described below.   By their nature, dynamic
methods can produce flows of mixtures of known composition for considerable periods
of time and so are particularly convenient sources of mixtures for testing sampling
techniques.

All dynamic methods involve mixing of flowing gases, and efficient and thorough
mixing within the time of residence of the gases in the mixing chamber is essential
for accuracy.   The chamber must have sufficient volume to give adequate residence
time at the flow rates which are to be used, and should be so designed that gases
do not readily have a direct path from the entrance to the exit.   Movement of the
contents must be induced within the chamber.   This can be done mechanically by
means of a stirrer, or can be induced by the shape of the vessel.   Stagnant volumes
develop readily in spherical vessels, and smooth flow without mixing takes place
in cylindrical vessels.   Turbulent flow with mixing can be generated by passing
the gases through small diameter tubing or by inserting baffles in a cylindrical
vessel.   A vessel which consists of alternate sections of large and small diameter
promotes mixing of flowing gases because of the changing flow pattern as the gases
move through it.   Such a vessel can conveniently be made by connecting a number of
small spherical vessels in series.   The diameter of the tubes which connect the
wider sections should not be so small as to impose any appreciable resistance to
flow on the gas stream.

In using dynamic methods, adsorption of certain components of the mixture may occur
on parts of the apparatus.   It is important, therefore, to run the mixing equipment
for some time, both on starting up and after changing the concentration of the mixtures which are being prepared, before the mixtures are used for calibration or test
purposes.   Equilibration between the gas stream and the surfaces of the equipment
may take from a few minutes to several hours depending on the nature of the substances in the gas mixture.

The following methods of dynamically preparing mixtures of known concentration have
given satisfactory results.

Serial dilution of a standard mixture.   Many of the standard mixtures of gases
which are commercially available in cylinders have a concentration of the required
substance which is too great for instrumental calibration or testing of analytical
methods.   However, a stream of the standard gas can be diluted dynamically by a

stream of air to give a mixture which has a lower concentration of the required
gas.   The flow of gas from the cylinder and the flow of air must be metered
accurately, either by calibrated gas flowmeters or by a gas metering pump.   The
metered gases must then be mixed in a vessel such as has previously been described.
The concentration of the required gas in the mixture leaving the mixing chamber is
its concentration in the standard gas divided by the ratio of the total flow to the
flow of standard gas.

Vapour saturation.   This method is suitable for making mixtures with air of the
vapour of a relatively involatile liquid.   Air is passed at a measured rate over,
or bubbled through, the liquid, which is in a vessel immersed in a constant temperature bath, and is then mixed with diluent air.   In theory, the partial pressure
of the vapour of the liquid in the air leaving the vessel should be the saturated
vapour pressure of the liquid at the temperature of the constant temperature bath.
In practice, saturation is seldom achieved in a simple device of this type due to
problems of heat transfer and to reduction of the temperature of the evaporating
liquid by the heat of evaporation.

The following method can be used to saturate air with a vapour at a given temperature.   The air is first presaturated at a temperature 3 to 5°C higher than the
final temperature.   This is achieved by bubbling the air through the liquid which
is contained in a series of vessels at the higher temperature.   The number of
vessels required may depend on the particular liquid being used, but, with three
vessels in series, over 90% saturation can usually be achieved.   The air is then
cooled to the temperature finally required, and is bubbled through another series
of vessels at this temperature which also contain the liquid.   The air is over-
saturated at this lower temperature and loses some of the vapour.   When it leaves
the vessels the air is 100% saturated at the lower temperature.   The saturated air
must be mixed with the diluent air without undergoing any change in temperature,
otherwise the vapour will condense in the gas lines and this will lead to error in
the final concentration.

The concentration of the vapour is calculated from the saturation vapour pressure
at the final temperature of saturation, and the ratio of the rates of flow of the
saturated and diluent air.

Motor driven syringe.   This technique is suitable for preparing dilute mixtures in
air of gases, or of the vapour of relatively volatile liquids.   The gas, the liquid,
or a solution of the liquid in an inert solvent which will not interfere in the
measurements which will be made on the mixture, is injected slowly and continuously
by a motor driven ram pump or syringe into a stream of air of known flow rate.   The
time over which the system can operate depends on the capacity of the pump cylinder
and the rate of flow of liquid from the pump.   Suitable pumps for this purpose can
be constructed in the laboratory.   The design of a highly versatile pump of very
high accuracy has been published (Ref. 8).

When the equipment is used to produce a dilute mixture of a gas in air, the gas
concerned is injected from the syringe through a capillary into a mixing vessel
where it is mixed with a measured flow of diluent air.   When used to add a component to air in liquid form, the syringe forces a slow stream of liquid through a
nozzle into a container of large volume in which evaporation and mixing takes place.
A suitable mixing vessel consists of an aspirator of 3 to 5 litres capacity, laid
on its side with the normal side outlet pointing vertically downwards.   An atomiser
nozzle enters the neck of the aspirator horizontally through a tight fitting bung,
and diluent air enters the vessel through the outer concentric tube of the atomiser.
The mixture is taken from the mixing vessel through the vertical outlet.

In operation, the motor of the pump and the flow of diluent air are started, but, because of the size of the mixing vessel, the device must be operated for some time, depending on the flow rates, to allow the concentration to reach a stable value before it can be used as a source of gas of constant concentration. The flow rate of diluent gas must be comparatively high, usually about 10 litres per minute, to achieve satisfactory evaporation and mixing, and so it may be necessary to provide a means for disposing safely of the unwanted gas. The concentration delivered by the device is calculated from the rate at which liquid is pumped by the ram, and the rate of flow of diluent air.

By injecting a solution of 1% of toluene di-isocyanate in toluene with this type of apparatus into an air stream of 10 litres per minute, concentrations of between 0.01 and 0·1 ppm in air have been produced. (Ref. 9).

Diffusion cell.   This device is based on the principle that the diffusion of a vapour through a tube is dependent only on the geometry of the tube, the nature of the gas and the temperature.   If the vapour at one end of the tube is produced continuously by the evaporation of liquid, the pressure of vapour at that end of the tube will be equal to the vapour pressure of the liquid at the temperature at which it is maintained.   If the other end of the tube is continuously swept by an air stream the pressure of vapour there is almost zero.   Under these conditions the rate of diffusion of vapour along the tube is given by the equation:

$$r = (DPMA/RTL)\ln(P/(P-p)) \qquad\qquad (1)$$

where r = rate of diffusion of vapour in moles/cm^2/s

D = diffusion coefficient of the vapour in cm^2/s

M = molecular weight of the vapour

P = total pressure

R = gas constant

T = absolute temperature

p = vapour pressure of liquid

A = cross sectional area of tube in cm^2

L = length of tube in cm.

The diffusion coefficients for many gases are recorded in the International Critical Tables (Ref. 10) at the standard conditions of 0°C and 1 atmosphere pressure.   The values at other temperatures and pressures can be calculated using the formula:

$$D = D_0 T P_0 / T_0 P \qquad\qquad (2)$$

where D = diffusion coefficient at temperature T and pressure P

D_0 = diffusion coefficient at temperature T_0 and pressure P_0

T_0 and P_0 are the standard temperature and pressure for which the data are recorded.

Combining equations 1 and 2:

$$r = (D_0 P_0 MAT/RT_0^2 L) \ln(P/(P-p))$$

Thus for a given cell and a given liquid the diffusion rate depends only on the total gas pressure and the vapour pressure of the liquid in the cell. If the total pressure is maintained constant the diffusion rate is a function of temperature only.

The design of a practical cell is described in Ref. 11 and is shown in outline in Figure 8.1

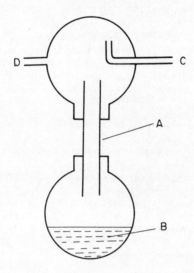

Fig. 8.1 Diffusion cell.   A = Diffusion tube; B = Liquid;
C = Diluent air inlet; D = Exit for gas mixture.

The flasks are attached to the diffusion tube by ground glass joints to allow the tube to be changed. The lower flask becomes saturated with vapour at the working temperature. The vapour diffuses along the straight tube, mixes with the diluent air stream in the upper flask and is carried away by it. The rate of diffusion of the vapour is calculated from the above equations, and the volumetric concentration of vapour in the air stream is calculated from this rate of diffusion and the air flow rate, using the gas laws and the molar volume of the vapour.

The temperature of the cell must be controlled accurately and a bath with a high quality thermostat is required. If the diffusion coefficient of the vapour concerned is not known the cell has to be calibrated. This can be done by measuring the concentration of gas leaving the device by an independent means or, alternatively, by measurement of the loss of weight of the lower flask. This type of calibration should always be carried out when the concentration of the gas is required to be known accurately, because the rate of diffusion may, in practice, deviate slightly from the theoretical value because of distortion of the tube during construction of the cell, and end effects due to the flow of air over the end of the tube.

The dimensions of the cell should be chosen specifically for each application. The cell should generate the appropriate amount of vapour to be mixed with a stream of air at the required flow rate to give the desired concentration, while working at a temperature appropriate to the liquid concerned. The air flow through the cell should be metered accurately. The concentration given by the cell can be varied by changing the air flow, the dimensions of the diffusion tube, or the working temperature.

Permeation cell. The low flow rate of gas or vapour which is to be diluted is produced by allowing the gas or vapour to permeate through a wall of plastics material under controlled conditions. (Refs. 12, 13, 14). In a common form of permeation cell, the volatile liquid, the vapour of which is required, is contained in a closed cylinder of the plastics material. The mechanism of permeation is complex and depends not only on diffusion, but also on the mutual solubilities of the liquid and the polymer. The following equation, however, has been suggested to describe the rate of permeation of a vapour through the walls of a cylinder, at a constant temperature:

$$F = 2\Pi DS \ (P_2 - P_1)/\ln(b/a) \tag{4}$$

where F = permeation flux per unit length

D = diffusion constant of liquid in polymer

S = solubility constant of liquid in polymer

P_2 = partial pressure of vapour outside cylinder

P_1 = partial pressure of vapour inside cylinder

$2a$ = inside diameter of cylinder

$2b$ = outside diameter of cylinder.

D and S are functions of temperature, the properties of the permeating gas, and of the material of construction of the cylinder. For a given vapour and cylinder material, the permeation rate varies with temperature, within a limited range, according to the relation:

$$\log \ (F_1/F_2) \ = \ -K(1/T_1 - 1/T_2)$$

where F_1 and F_2 = permeation rates at absolute temperatures T_1 and T_2 respectively

K is a constant.

These relations can be used to give approximate dimensions of a tube for particular applications, but cannot normally be used for accurate design purposes because the constants involved are rarely known with accuracy. Further, they take no account of deviations from the theoretical value which occur in practice, for example, as a

result of leakage past the seals.   The optimum design of tube for any specific
purpose is arrived at by experiment.

The most commonly used material for the construction of permeation tubes is polytetrafluoroethylene, but fluorinated ethylene-propylene resin has been used to make
tubes which have a lower permeation rate.   These polymers are easily worked, and
are available in the form of tubes of a variety of diameters and wall thicknesses.
When equilibrium in permeation has been attained the rate is almost independent of
the wall thickness, and this factor can be varied to take account of the different
pressures within the tubes when different liquids are introduced.   The polymers
are inert and have low solubility in most organic compounds.

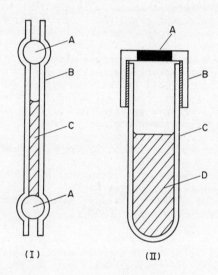

Fig. 8.2   (I) Permeation tube. A = steel ball; B = permeable
tube;   C = liquid source of vapour.   (II) Permeation
capsule. A = plug of permeable plastics material;
B = impermeable cap;   C = impermeable tube;
D = liquid source of vapour.

A permeation tube can be made by sealing a length of plastics tubing at both ends.
The seal is conveniently made by forcing into each end, a steel ball which has a
diameter of about 1·5 times the internal diameter of the plastics tube. (Fig. 8.2).
The seal can be opened sufficiently to allow the tube to be filled from a source of
the liquid of vapour under pressure by pinching the tube around one of the balls
with a pair of rubber-padded pliers.   Volatile liquids can be condensed within the
tube if it is immersed in a cold bath.

This permeation tube has a comparatively large area of permeable surface and so has
a high permeation rate.   If this rate is too great for a particular application, a
permeation capsule with a smaller exposed area of polymer can be made from a tube
of metal or glass which is permanently sealed at one end and closed at the other end

with a special screwed cap.   The cap can either be made of the permeable polymer, or of an impermeable material which has a plug of the polymer inset into it. (Fig. 8.2.).   In a tube of this construction the exposed area of permeation surface can be made very small, and very low permeation rates can be achieved.   The weight of the capsule should be kept as small as possible.   In making any permeation tube or capsule great care should be taken to avoid heating the plastics material otherwise its physical properties may be changed.   The change may alter the permeation rate which will return to its normal value only very slowly.

The rate of permeation from such a tube or capsule is very dependent on the temperature.   In use, the tube or capsule is contained in a double walled vessel, and water from a constant temperature bath is circulated between the walls.   The temperature of the water must be controlled accurately by a high quality thermostat. After some time, which may vary from several hours to several days, the rate of permeation reaches a constant value and maintains that value until the supply of vapour within the tube is almost exhausted.   The rate of permeation is measured by weighing the tube or capsule at intervals.

The diluent air is flowed at a measured rate through the constant temperature enclosure which contains the permeation tube.   The volumetric concentration of the vapour in the air leaving the cell is calculated from the permeation rate, the flow rate of the diluent air, the gas laws, and the molar volume of the vapour concerned. This concentration can be changed by altering the flow rate of the air or by changing the temperature at which permeation takes place.

However, temperatures around the transition temperatures of the polymer which forms the permeation wall should be avoided.   The physical changes which take place in the polymer at these temperatures may cause instability in the permeation rate. The accuracy of the concentration of the mixture produced by a permeation cell depends on a number of factors.   The temperature must be controlled accurately and should be maintained constant within $0.05^{\circ}C$ to ensure that the permeation rate varies by less than 1%.

The weight of the permeation tube or capsule must be determined to high accuracy to calculate the weight loss.   A typical daily weight loss for a plastics tube device is a few milligrams, and the glass or metal capsule device may lose one hundredth of that amount.   A very sensitive analytical balance is required to make this measurement, and weights should be determined at constant pressure and humidity. This is the factor which is most likely to determine the accuracy with which the concentration of the gas is known.

The flow rate of the diluent air must be measured accurately by a calibrated flowmeter.

If all precautions are taken, this device can produce, for long periods, a mixture of which the composition is constant and accurately known.   It has been reported that under favourable conditions a stream of air which contains 1 ppm of vinyl chloride can be produced with ± 2% with a standard deviation of less than 0.05 ppm. (Ref. 15. Analytical Note AN 22).

PREPARATION OF STANDARDS IN LIQUID SOLUTION

Although in atmospheric monitoring the samples are taken in the gas phase, some of the analytical devices make measurements on the required substance when it is in solution in a suitable solvent.   For calibrating such instruments solutions of the substance concerned at known concentration in an appropriate solvent are normally required, though certain analytical devices can be calibrated by either gaseous or liquid samples.   For example it has been shown (Ref. 16) that the gas chromatographic peak produced by vinyl chloride from a gas sample was not significantly different from the equivalent peak produced by a solution of vinyl chloride in an organic solvent.

Solutions of known concentrations of involatile substances in organic solvents or water can be made by standard gravimetric or volumetric techniques, and these may be serially diluted to prepare solutions of very low concentration.   Volatile substances, vapours and gases are more difficult to handle and great care is necessary to avoid loss of volatile components of the mixture.   If the volatile substance is not highly soluble in the solvent it may concentrate in the vapour space within the vessel.   Under these circumstances the vessel should be kept as full as possible to reduce the vapour space.

The following methods have been found to be satisfactory for the preparation of solutions of volatile substances of known concentration.

Volumetric Method

This technique is suitable for making mixtures of a vapour in a solvent in which it is very soluble, and was developed by NIOSH for making standard solutions of vinyl chloride in carbon disulphide. (Ref. 16).   1 ml of the vapour is taken in a gas syringe by one of the methods previously described.   The needle of the syringe is inserted under the surface of about 5 ml of the solvent which is contained in a 10 ml volumetric flask.   The plunger of the syringe is pulled out a little to draw solvent into the syringe.   This solvent dissolves vapour in the syringe, and draws more solvent into the barrel until the vapour is completely dissolved.   The solution is ejected into the volumetric flask and the syringe is rinsed with fresh solvent which is added to the flask.   The flask is then filled to the mark with solvent and the contents are thoroughly mixed.   The concentration in gravimetric terms is calculated from the vapour density of the vapour.   The solution may be further diluted as required.

Gravimetric Methods

Gases and vapours may be absorbed in solvents by bubbling them through the liquid. The amount dissolved may be determined by weighing and, from the weight addition, the concentration of the solution may be calculated.   However, if the concentration required is low the weight added is very small, and to enable the vessel to be weighed accurately on an analytical balance the weight of the vessel must be kept to a minimum.   Further, the gas flow must be kept to a value low enough to avoid loss of solvent by evaporation or entrainment in any gas which does not dissolve but passes through the absorber.

Flask method.   NIOSH (Ref. 17) uses a technique of this type for preparing standard solutions of vinyl chloride gravimetrically.   Vinyl chloride is bubbled slowly into a weighed 10 ml volumetric flask which contains about 5 ml of toluene.   When a weight addition of between 100 and 300 mg has been observed the solution is diluted to exactly 10 ml with carbon disulphide.   This solution is then diluted with carbon disulphide to provide solutions of the required concentrations.

Sealed cell method.   An alternative type of vessel in which the absorption may be carried out is shown in Fig. 8.3.   This vessel, which has a capacity of about 15 ml is sealed to prevent loss of solvent, and aliquots for further dilution can be withdrawn through the septum.

Capillary tube method. An entirely different gravimetric method consists of adding
a known weight of the substance which is to be dissolved, to a known weight or volume of solvent.   The substance, in the liquid phase, is added in a sealed tube.
A weighed glass capillary tube, sealed at one end, is cooled in solid carbon dioxide,
either by clamping it between two pieces of refrigerant or by inserting it in a
small hole drilled in a block of the refrigerant.   The vapour which is to be diluted
is flowed into the capillary through a hypodermic needle and condenses.   When
sufficient liquid is present in the capillary, the needle is removed and the glass

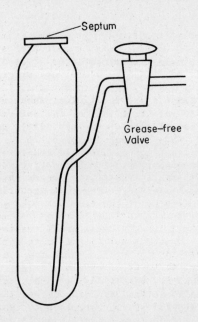

Fig. 8.3   Sealed vessel for gravimetric preparation of
standard mixtures.

capillary is sealed while it is still refrigerated, using a small flame.   The
sealed capillary is weighed to determine the weight of liquid which it contains,
and the weight is checked at intervals to ensure that there is no leakage of vapour.
The capillary is scratched in several places to assist in breakage and is placed in,
and resting against the side of, a flask which contains a known volume of solvent.
The flask is closed by a rubber stopper through which passes a metal rod, the end
of which has a V-notch cut in it, and which reaches into the flask.   By manipulation
of this rod the capillary is broken and its contents are dissolved in the solvent.

This technique, and a modification which uses a smaller vessel in which the capillary is broken by a mechanism which is contained within the vessel, have been used
successfully for preparing dilute solutions of vinyl chloride in various organic
solvents and in water.   (Ref. 15, Standard Methods SM1 and SM18).

TESTING OF MONITORING METHODS

Flows of gases of known composition from cylinders, or produced by the dynamic methods described above, are suitable for testing and evaluating methods of sampling and measurement.   The gas flows can be maintained for long periods with negligible change in composition, and the concentration of the substance which is being monitored can be readily changed.   However, tests carried out in the laboratory using such gas streams are not completely representative of the measurements which have to be made in practice.   For example, the laboratory source usually provides a known concentration of the substance concerned in clean, dry air.   In practice, other contaminants are usually present, and the humidity of the sample may vary considerably.   The effect of humidity on the procedure can be determined in the laboratory by using test gases prepared with humidified air.   This can be done by saturating with water vapour all or part of the diluent air at a known temperature, using the vapour saturator previously described or, alternatively, by using a simple but less efficient saturator and measuring the resulting humidity of the air leaving it.

The effects of other atmospheric contaminants on the monitoring procedure can be checked by testing the method on a mixture which contains these contaminants in addition to the substance which is being measured.   Such a mixture can be made by a number of methods.   For example, a number of gas metering and mixing pumps can be used in parallel, each handling a separate substance.   Alternatively, a number of permeation tubes, each filled with a different substance, can be used in series in the same constant temperature enclosure.   If, however, the various tubes require different temperatures, they must be housed in separate enclosures and the effluent gases mixed in the correct proportion.

Most laboratory tests of monitoring methods are carried out on gaseous mixtures of constant composition.   However, the concentration of gases in the workplace changes from time to time, and it is for this reason that measurements of time-weighted average concentrations are required.   Methods for making these measurements must be tested over the range of concentrations which is expected, but even that does not show the response of the method to a dynamic situation of changing concentration.   The laboratory test which is normally used to evaluate a method under conditions of changing concentration is one in which the concentration is changed from one to another of two values at known intervals of time.   Although this does not truly simulate practical conditions it allows the theoretical value of the average concentration to be calculated, and also tests some of the factors which are likely to affect the response of the method to changes in concentration.

Whenever possible methods should be evaluated in the actual conditions under which they are to be used.   If two or more different methods are available from which results can be obtained, they should be used in parallel and the results compared. If the methods are very different and use different methods of sampling, if they make measurements over different times, or if they are differently affected by other atmospheric components, due allowance for these factors must be made when the results are compared.   If, however, two or more methods of sampling and measurement show very similar results there is a high probability that the results are true.

The difficulties of conducting such a comparison must not be underestimated.   If two samples are taken from adjacent points by the same sampling procedure and are analysed by the same method, they will often give results which differ by an amount which exceeds the uncertainty of the analytical procedure.   Within plants which use certain types of processes, and in the environment surrounding them, the concentration of atmospheric pollutants at any point resulting from the operation of the plant varies rapidly with time.   The type of change which can be expected is illustrated in Fig. 8.4 which is a record of the output from a continuous analyser which

was monitoring the concentration of a chlorinated hydrocarbon at a point in such a
plant.   The concentration may also vary significantly over short distances, depending on the nature of the source of the contamination and the pattern of air flow
within the plant.   In these circumstances it becomes difficult to ensure that
samples taken over the same time, even from adjacent points, are strictly comparable.
This is especially true of techniques which measure instantaneous or short term
average concentrations, and under these conditions the comparison can be made only
on a statistical basis.

The importance of testing the performance of monitoring methods and establishing
the accuracy of measurement which is possible will be apparent from the next chapter.

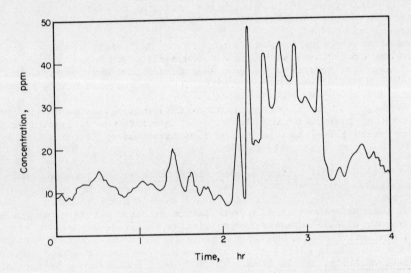

Fig. 8.4   Variation in concentration of a chlorinated
hydrocarbon at a point in a chemical plant.

REFERENCES

1.   Air Products Ltd., Special Products Dept., Crewe, Cheshire, U.K.

2.   British Oxygen Co. Ltd., Special Gas Division, Morden, London, U.K.

3.   Rank Precision Industries, Margate, Kent, U.K.

4.   EDT Supplies Ltd., London, U.K.

5.   K. Leichnitz, Calibration methods for Draeger tubes, Draeger Review, 41,
     12 (1978).

6.   Gas Analyser and Monitor.   Foxborough Analytical Limited, Milton Keynes,

Bucks, U.K.

7. Calibration Gas Blender.   CR Laboratory Services, Worthing, Sussex, U.K.

8. W.E.A. Ruska, G.F. Carruth, & R. Kobayashi, Micropump - an apparatus for
    steady state synthesis of gas mixtures at very dilute concentrations,
    Rev. Sci. Inst. 43, 1331 (1972).

9. The Preparation of Known Concentrations of Toluene Di-isocyanate.   Internal
    Laboratory Report, Universal Environmental Instruments, Poole, Dorset, U.K.

10. International Critical Tables Vol. V, McGraw Hill, New York (1929).

11. J.M. McKelvey & H.E. Hoelscher, Apparatus for preparation of very dilute
    gas mixtures, Anal. Chem. 29, 123 (1957).

12. A.E. O'Keefe & G.C. Ortman, Primary standards for trace gas analysis,
    Anal. Chem. 38, 760 (1966).

13. A. Credergren & S.A. Fredriksson, Trace analysis for chlorinated hydrocarbons
    in air by quantitative combustion and coulometric chloride determination,
    Talanta, 23, 217 (1976).

14. D.A. Ferguson & K.J. Saunders, Quantitative Calibration Procedures and Atmos-
    pheric Sampling in Environmental Analysis.   International Conference on the
    Monitoring of Hazardous Gases in the Working Environment, London, U.K. Dec.
    1977.

15. W. Thain (Edit).   The Determination of Vinyl Chloride.   Chem. Ind. Assoc.
    Ltd., London. 3rd Ed. (1977).

16. R.H. Hill, C.S. McCammon, A.T. Saalwaechter, A.W. Teass & W.J. Woodfin, Gas
    chromatographic determination of vinyl chloride in air samples collected on
    charcoal, Anal. Chem. 48, 1395 (1976).

17. NIOSH Manual of Analytical Methods.   Method No. P & CAM 178, U.S. National
    Institute for Occupational Safety and Health, Cincinnati, Ohio, U.S.A. (1977).

Chapter 9

STATISTICS OF MONITORING

INTRODUCTION

The interpretation of the results of monitoring is simplified and assisted by an understanding of the variables and variations which exist in the measured quantities and the methods of measurement. For this an appreciation of the methods of statistical analysis is required. Most analysts are familiar with statistical techniques, but some of the problems of environmental analysis require concepts beyond those normally employed in the analytical laboratory. These are outlined in this chapter following a brief introduction to the methods more commonly applied, which is given for the sake of completeness and continuity. No attempt is made to justify the methods nor to derive the relations used. This information can be found in text books on statistics.

STATISTICAL CONCEPT OF ERROR

Every measurement is subject to some uncertainty. It is unfortunate that the term "error" is used to describe the uncertainty because, at least to the uninitiated, it suggests negligence or lack of expertise on the part of the operator. However, it is used merely to describe the results of imperfections in the equipment or method of measurement, and the uncertainty which accompanies its use. Errors can be of two types, systematic and random. Systematic errors arise from causes such as imperfection of equipment, for example, an error in calibration of a volumetric flask. The flask when used will always contain the same volume but this will be different from what the user expects it to be. Random errors arise from the difficulty of using equipment, for example in making up the volume in the flask exactly to the calibration mark. Measuring equipment is designed to minimise the random errors associated with its use, and measuring equipment for the fundamental properties of weight, volume, time, etc. show little random error when used by experienced operators. Random errors can be introduced by poor laboratory practice. Test methods should be adequately described and strictly followed to eliminate this source of random errors.

Precision and Accuracy

The magnitude of error is expressed by two statistical concepts, precision and accuracy. Precision is a measure of the scatter of results obtained when a

measurement is made repeatedly on the same sample.   Accuracy is a measure of the
difference between the true value and the value given by measurement.   Thus,
accuracy indicates the magnitude of the systematic errors and precision indicates
the magnitude of random errors.   The ultimate goal in the development of analytical
methods is to find a method of measurement which is both accurate and precise.

Systematic Errors

Systematic errors can arise from inadequate standardisation of measuring equipment,
for example, volumetric vessels and weights, improperly calibrated instruments,
physical limitations and chemical interferences in procedures, which lead to errors
such as incomplete recovery of substances from mixtures or solutions.   Systematic
errors can also arise from personal bias, for example in observation, particularly
of colour, in handling of equipment, and consistent errors in calculation.

Systematic errors are themselves of two types.   Additive errors have a constant
value which is independent of the magnitude of the quantity which is being measured.
Proportional errors depend in magnitude on the amount of the measured quantity.
The type of error can be identified by making measurements at known different levels
of the measured quantity.   If the measured value is plotted against the true value,
the line may not in either case pass through the origin, but will be straight if
the error is additive, and curved if the error is proportional.   The magnitude of
the error can be calculated from the axis intercept in the first case and from the
curvature in the second.   Such tests require a source of materials for which the
true value of the property which is being measured is known.   In the case of
monitoring methods these can be mixtures prepared by the methods described in
Chapter 8.

When the magnitude and type of systematic errors have been found, the sources of
error should, where possible, be identified and eliminated.   It is possible to
compensate for a systematic error by applying an empirical correction.   However,
this should be done only if the cause of the error is understood.   For example,
if a systematic error is due to imperfect calibration of a piece of equipment,
that error will be found every time that specific piece of equipment is used.   If,
however, the same method is used with an identical piece of equipment which is
perfectly calibrated, the error would not be present, and any correction for the
original error which is included in the method will be unnecessary.   Correction
procedures for systematic error are correctly applied in the NIOSH method for
recovering substances from charcoal traps by solvent desorption.   (See Chapter 6).
It is recognised that the desorption process is not complete and a correction factor
is applied in the calculation to allow for the incomplete recovery.   The correction
factor, however, must be determined for each substance by each laboratory to allow
for slight differences in procedure, and also for each batch of charcoal to allow
for differences in physical properties.

Where systematic errors are eliminated by compensation methods the correction should
be reviewed if the method is altered in any way which may change the magnitude of
the error.   Internal standards are often used to compensate for systematic errors.

Systematic errors become increasingly significant as the magnitude of the measured
value decreases.   Eventually they place a limit on the quantity which can be
detected, and when the error is of about the same magnitude as the measurement the
method becomes unreliable.

Analytical methods should not be extrapolated beyond the ranges in which they have
been proved.   Many of the physical and chemical processes involved in analysis
have a linear response only over a limited range, though they may continue to

respond outside that range.   This introduces systematic errors which would not
have been found during the evaluation of the method.

Random Errors - The Normal Distribution

Random errors give rise to deviations from a true value and are as often positive
as negative.   The error may take any value, and if a large number of measurements
are made of a fixed quantity and averaged, the random errors will tend to compensate
for each other.   Thus, the more measurements which are made and averaged the nearer
the average will be to the true value.

If a measurement which is subject only to random errors is repeated many times, and
the number of occasions on which each value of the result is obtained is plotted
against that value, the curve obtained is that of the normal distribution. (Fig. 9.1).

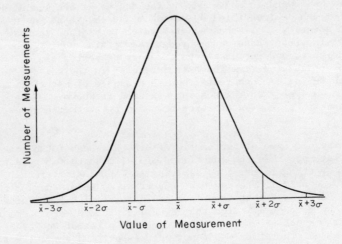

Fig. 9.1.   The normal distribution of values of a measurement
subject to random error.

Very large numbers of observations are required to produce this curve accurately,
but smaller numbers show the same trend as the curve is built up.   The curve is
symmetrical about a value which is the arithmetic mean of all the measurements and,
theoretically, continues asymptotically to zero in both directions.   If sufficient
measurements are made the arithmetic mean will equal the true value of the property
which is measured.

The scatter of the values of measurement is indicated by the width of the curve and
so a means of describing this is required.   The curve can be described completely
by two parameters, the arithmetic mean which gives the value at the peak, and the
standard deviation which is a measure of the width.   These are defined by the
relations:

$$\bar{x} = (\Sigma x)/n \tag{1}$$

$$\sigma^2 = (\Sigma(\bar{x} - x)^2)/n \qquad\qquad (2)$$

where Σx = sum of all observations

$\bar{x}$ = arithmetic mean

$\Sigma(\bar{x} - x)^2$ = sum of the squares of the differences of all observations from the arithmetic mean

σ = standard deviation

n = number of observations.

It will be seen that equation (2) for evaluating the standard deviation involves lengthy calculation, and another form which is easier to manipulate is:

$$\sigma^2 = (\Sigma x^2 - (\Sigma x)^2/n)/n \qquad\qquad (3)$$

The standard deviation is sometimes expressed in terms of two other parameters:

Variance = σ^2

Coefficient of Variation = $\sigma/\bar{x}$

or, expressed as a percentage, $100\sigma/\bar{x}$

The Coefficient of Variation is sometimes called the Relative Standard Deviation.

The value of σ defines the general shape of the curve.   In particular if the curve is divided by vertical lines at $\bar{x}$ and spaced at distances equal to σ above and below that value (see Fig. 9.1) the areas under the curve lying between the lines are:

Boundaries	% of total area which lies between boundaries
$\bar{x} + \sigma$ and $\bar{x} - \sigma$	68.3
$\bar{x} + 2\sigma$ and $\bar{x} - 2\sigma$	95.5
$\bar{x} + 3\sigma$ and $\bar{x} - 3\sigma$	99.7

Thus if a single measurement is made of this quantity there is a 68.3% chance that it will deviate from the arithmetic mean or true value by less than σ, a 95.5% chance that it will deviate by less than 2σ, and a 99.7% chance that it will deviate by less than 3o.

The value of σ for a monitoring method can be evaluated from repeated measurements
of a mixture of known concentration.   The results will show a scatter, and if
plotted as indicated above, will give a curve which usually approximates to the
normal distribution.   The arithmetic mean and standard deviation can be calculated
from equations (1) and (2) or (3).   If the arithmetic mean is equal to the known
value of the concentration there is no systematic error.   If not, the difference
between the two values is a measure of the systematic error.   The standard deviation
is a measure of precision or random error.

Provided the test mixture has a constant composition over the time during which the
tests are being carried out, the value of the standard deviation obtained will be
reliable.   However, the value for the systematic error depends on the test mixture
having a concentration which is known with accuracy.   The preparation of a test
mixture is subject to experimental error, as is the analytical method being tested.
Therefore, great care should be taken in the preparation of test mixtures to minimise
the errors, and it is for this reason that it was specified in the previous chapter
that the methods should be capable of independent calibration, preferably by
gravimetry.

The values should be checked to ensure that they do comply with the normal distribution before analysis.   If they do not, the method should be examined for unusual
errors, for example cyclical errors which occur at definite times, and these should
be eliminated.

The causes of systematic error should be eliminated or a compensation for the error
introduced as previously described.   The standard deviation is then a measure of
the uncertainty which remains in the measurement when it is made by the method
which has been tested.

The overall standard deviation of the monitoring procedure depends on the standard
deviations of all the processes which have been involved in the procedure.   For
example if the procedure involves three stages

        Stage 1   sample collection

        Stage 2   sample recovery

        Stage 3   measurement

each will have associated with it a standard deviation.   Then, provided that the
errors at the three stages are uncorrelated, the standard deviation of the total
procedure is linked to the other standard deviations by the relation:

$$(\sigma \text{ of total procedure})^2 \;=\; (\sigma \text{ of stage 1})^2 \;+\; (\sigma \text{ of stage 2})^2 \;+\; (\sigma \text{ of stage 3})^2$$

On occasions it is useful to evaluate the standard deviations of the individual
stages in the study of methods.

It is possible to calculate from the properties of the normal distribution the
probability that any particular value of the variable will not be exceeded.   This
function of cumulative probability is the basis of a graphical method of linear-
ising a normal distribution, analogous to the use of logarithmic graph paper to
linearise an exponential function.   Arithmetic probability paper is printed in
such a way that the ordinate is proportional to the cumulative probability function
and the abscissae are marked linearly.   The results of measurements are then

plotted on the probability paper in the following manner.   The observations are
arranged in ascending order of magnitude and numbered 1 to n.   These form n + 1
intervals in the total distribution which extends from $-\infty$ to $+\infty$.   The probability
of a particular value lying in a particular interval can be estimated and the
percentage cumulative probability functions for the various observations can be
calculated. (See Table 9.2, page 146).   The values of the observations are then
plotted on the arithmetic probability paper against the value of the appropriate
cumulative probability.   If the observations are from a normal distribution the
points plotted will lie on a straight line.   The value of the mean, $\bar{x}$, is that
corresponding to a cumulative probability of 50%, and the value of $\bar{x} + \sigma$ is that
corresponding to a cumulative probability of 84.13%.

Confidence Limits

If the true value of a quantity is X, and it is measured a very large number of
times by a method which is subject only to random errors, the measurements will
have an arithmetic mean equal to X.   If, however, the quantity is measured by the
same method a lesser number of times n, and the arithmetic mean $\bar{x}$ and standard
deviation σ are calculated, there will be some uncertainty about the values of
these quantities, because of the limited data.   However, it can be stated with
95% confidence that the true value X will be within the range $\bar{x} \pm 2\sigma/n^{0.5}$.   These
values are known as the Upper and Lower 95% Confidence Limits of the Mean.   Limits
for other degrees of confidence can be evaluated but in monitoring applications
the 95% limit is most commonly used.

This expression for confidence limits of the mean holds for numbers of observations
of about 20 or more.   If the number of observations n is less than 20, a better
approximation to the standard deviation which would have been obtained from a
large number of observations, can be obtained from the small number by using the
formula:

$$S^2 \;=\; (\Sigma(\bar{x} - x)^2)/(n - 1) \tag{4}$$

or using the alternative form

$$S^2 \;=\; (\Sigma x^2 - (\Sigma x)^2/n)/(n - 1) \tag{5}$$

where   S   = best estimate of the standard deviation

This is then used to calculate the 95% confidence range from the relation:

$$95\% \text{ confidence range } \;=\; x \stackrel{+}{_{-}} t_{n-1}\; S/n^{0.5} \tag{6}$$

where t_{n-1} = the value of the Student t distribution for 95%
confidence, expressed in terms of number of
measurements n or number of degrees of freedom
n - 1.

The values of t are given in Table 9.1 for a number of values of n.   It will be seen that when only 2 or 3 measurements are made the confidence range is very wide, but that an increase in the number of observations beyond 10 does not greatly improve the confidence range of the mean.

TABLE 9.1 Values of the Student t distribution
for 95% confidence range

No. of Measurements n	Degrees of Freedom n − 1	Value of t
2	1	12.71
3	2	4.30
4	3	3.18
5	4	2.78
6	5	2.57
7	6	2.45
8	7	2.36
9	8	2.31
10	9	2.26
11	10	2.23
21	20	2.09
31	30	2.04
51	50	2.01
101	100	1.98

THE VARIATION IN CONCENTRATION OF CONTAMINANTS

The concentrations of atmospheric pollutants are not constant but vary continuously with time. The existence of pollutant gases in the atmosphere, both in the work-place and in the external environment, probably arises from two causes which are different in nature. There is a very small, but relatively constant, source of emission which arises from minute leaks in equipment, and there are random escapes which arise from irregular operations, plant malfunctions and accidental causes. The atmospheric concentration of the pollutant is, therefore, likely to be at a low and slightly varying level for most of the time, with peaks of higher concentration superimposed. The plant will have been designed and operated in such a way that escapes of the pollutant will occur infrequently, and the higher the concentration which is likely to result from an escape, the greater will have been the effort made to reduce its frequency of occurrence. The distribution of concentrations with time is, therefore, unsymmetrical. Further, the value of concentration cannot fall below zero and the high values arise with only low probability.

The occurrence of concentration values in a pattern of this type can be described
mathematically by the log-normal distribution.   It has been shown by a number of
workers that the variation of concentrations with time of a number of industrial
pollutants including dust, radioactivity and chemicals, both in working areas and
the external environment, is described reasonably well by this law. (Refs. 1, 2, 3,
4).   It is, however, difficult to demonstrate accurately that a series of measurements conforms to a log-normal distribution.   To define the lower part of the curve
satisfactorily it is necessary to make measurements to a high degree of accuracy,
and in pollution studies these values may be near the limit of detection of the
method of measurement.   There is also a problem in defining the curve at high
values of concentration because of the low frequency of occurrence.   However, the
distribution forms a useful model to describe the probability of occurrence of
concentrations of any particular value.

The Log-normal Distribution

The curve which shows the probability of occurrence of concentrations when defined
by this distribution is shown in Fig. 9.2.   The curve is skewed and in consequence
the mode, (the value which occurs most frequently), and the arithmetic mean do not
coincide.

If the values of the logarithms of concentrations which are log-normally distributed
are plotted against probability of occurrence, the distribution obtained is normal.
The parameters of the log-normal distribution can then be calculated from the
properties of this normal distribution.   Thus, if the value x represents a typical
value on the log-normal distribution and if

$$y = \log x \tag{7}$$

the values of y will be normally distributed and so

$$\bar{y} = (\Sigma y)/n \tag{8}$$

and

$$\sigma^2 = (\Sigma (\bar{y} - y)^2)/n \tag{9}$$

where $\bar{y}$ and σ represent the arithmetic mean and the standard deviation respectively
of the normal distributions of the logarithms.   Then when the antilogarithms of
these values are taken the following parameters of the log-normal distribution are
found:

$$\text{Geometric mean (G.M.)} = \text{antilog } \bar{y} \tag{10}$$

$$\text{Geometric standard deviation (G.S.D.)} = \text{antilog } \sigma \tag{11}$$

It can also be shown that the arithmetic mean, $\bar{x}$, of the log-normal distribution is
related to both of these quantities by the relation:

$$\log \bar{x} = \log (G.M.) + 0 \cdot 5 (G.S.D.)^2 \tag{12}$$

Thus, the arithmetic mean of a log-normal distribution is always greater than the geometric mean.

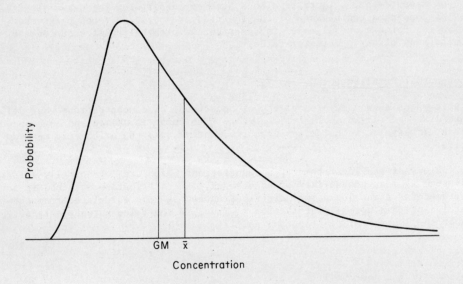

Fig. 9.2   The log-normal distribution of probability of
            occurrence of concentrations of an atmospheric
            pollutant.

If the value of n is small the best estimate of the standard deviation of the group of measurements should be calculated from equations (4) or (5) to correct for the bias found in analysing small groups.   Then:

$$s^2 = (\Sigma(\bar{y} - y)^2)/(n - 1) \tag{13}$$

$$\text{or} \quad s^2 = (\Sigma y^2 - (\Sigma y)^2/n)/(n - 1) \tag{14}$$

and the best estimate of the geometric standard deviation is:

$$G.S.D. = \text{antilog } S \tag{15}$$

The parameters of a log-normal distribution can be found graphically by the method described for graphical analysis of the normal distribution.   The cumulative percentage distributions are evaluated as previously described and these are plotted against the values of the relevant observation on logarithmic probability paper, or against the logarithms of these values on arithmetic probability paper. If the values belong to a log-normal distribution the points will lie on a straight line. If so, the values of log x will be normally distributed and the parameters of the distribution can be estimated.

APPLICATION OF STATISTICAL RELATIONS IN MONITORING

Monitoring programmes may be designed to determine whether regulations relating to exposure of workers and neighbours to atmospheric contaminants are being complied with or not. They may also be required to decide whether changes in process operations are having any significant effect on the level of atmospheric contamination. Both types of programme require consistent methods of measurement and means of comparing results, either with each other or with limits imposed by regulations. If the concepts of statistical variation in the concentrations and error in the measurements are accepted, the comparisons cannot be made on a basis of certainty, but can be expressed only as a degree of confidence. In monitoring work it has become customary to work to a confidence of 95%. The following examples show how statistical methods are used in making these comparisons.

Comparison of a Time-weighted Average Measurement with a Limit

A measurement of the time-weighted average concentration may have to be compared with a specified limit value to determine whether a worker's exposure has exceeded the prescribed limit. Only one value of the measurement is usually available, but the errors in the method can be assumed to be normally distributed and the statistical parameters of the method of measurement are available from previous testing. Then the best estimate of the time-weighted average concentration is that obtained directly from the measurement. However, due to the known errors in the measurement, the true value of the concentration may be different and it can be assumed, with 95% confidence, to lie within the range $\pm 2\sigma$ of the measured value, where σ is the known standard deviation of the method of measurement.

For comparison with a specified limit value it is necessary to determine where the limit value falls within this range. If the value of the best estimate is c, the probability of any other value occurring is shown in Fig. 9.3 where the curve is that of the normal distribution and the shape of the curve is determined by σ. Of particular interest is the value c', which is a lower value than c and which is located in such a way that 95% of the area of the curve lies above the ordinate at c'. There is then a 95% probability that the true value of the concentration has a value greater than c'. Thus, if the specified limit has a value of c' or less there is a 95% probability that the specified limit has been exceeded. The value of c' is given by the relation:

$$c' = c - 1.645\,\sigma \qquad\qquad (16)$$

Thus if

$$\text{Measured value} > \text{limit value} + 1.645\,\sigma$$

there is 95% probability that the limit has been exceeded.

This calculation assumes that the measurement has been made and averaged over the whole time specified in the limit.   If the measurement has been made over a shorter time the interpretation of the result depends on the assumption made regarding the concentration which existed during the remainder of the period specified in the limit value.   If the specified limit is expressed as an average over a time T, and it is found that the average over a shorter time, t, is c, with 95% confidence, then the minimum value of the average over time T is ct/T.   This assumes that the concentration over the remainder of the period, when measurement was not being made, was zero.   Only if this value exceeds the limit can there be 95% confidence that the limit has been exceeded.   However, unless there is evidence to the contrary, it is more realistic to assume that the average over the remainder of the time was the same as that during the time when measurement was being made.   In this case the time-weighted average would have been c if monitoring had continued for the whole time.

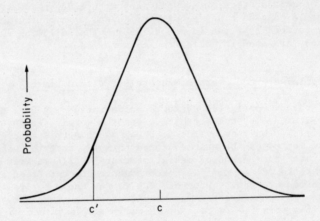

Fig. 9.3 The normal curve, showing value which is
exceeded with a probability of 95%

Analysis of Data from Random Measurements of Instantaneous Concentration

It may be required to determine whether a limit has been exceeded on the basis of a small number of determinations of instantaneous concentrations or values measured over very short periods of time.   The measured values will vary amongst themselves. Part of the variation will be due to errors in measurement, but these are likely to be small compared with the variation in atmospheric concentration of the contaminant from one measurement to another.   Thus, the values of the measurements are best described by a log-normal distribution.

The geometric mean and geometric standard deviation of the series of values can be calculated from equations (7), (8), (13), (14), (15) and the 95% confidence range of the geometric mean can be evaluated from equation (6) as follows:

    95% confidence range of the geometric mean

    $= \text{antilog } (\bar{y} \pm t_{n-1} S/n^{0.5})$

However, most limits are specified in terms of the arithmetic mean rather than the geometric mean. An exception is the Japanese Regulation for the limit of exposure of workers to vinyl chloride in which the limit is expressed in terms of the geometric mean and standard deviation of the monitoring measurements. The results of measurement as analysed above cannot, therefore, be used in comparison with regulatory limits. They can, however, be used as a means of comparing measurements of the same situation on a day to day basis, to determine trends and differences due to changes in operating procedures of the processes which are the sources of the contaminants.

The arithmetic mean of the measurements can, of course, be calculated for comparison with the limit value, but the confidence limits of this mean cannot be evaluated by any simple procedure. However, a method has been described (Ref. 2) for evaluating the measurements of benzene concentrations which are assumed to be log-normally distributed. The data are plotted on log-normal probability paper using Larsen's convention (Ref. 5). A line of limiting exposure is constructed on the same plot, and compliance or non-compliance is decided by the relation of the two lines.

Another technique for evaluating average concentration of a contaminant from grab sampling data and deciding whether a limit has been exceeded has been published. (Ref. 3). The confidence limits of the arithmetic mean is not calculated directly, but the geometric mean and the geometric standard deviation are evaluated, and are plotted on a special decision chart which shows boundaries between areas which indicate non-compliance, no decision, and no action. The boundaries between the areas are functions of the number of measurements. For such computations from grab sample data to be meaningful the distribution of values must be log-normal, and the samples must be statistically independent. This independence can be achieved if the samples are taken at random times, preferably selected by a truly random process, for example by a table of random numbers. Samples may be taken at regular intervals if the concentration values vary randomly and fluctuations are of short duration compared with the time over which the average is specified. However, there is always the possibility that samples taken at regular intervals may show bias because of cyclical variations in the process which is the source of the contamination.

An exception to random time sampling is used when the samples are required for checking compliance with a ceiling value. These samples should be taken when the concentration is expected to be at its greatest value.

A method of confirming by random measurements that the values of the concentration of a contaminant are log-normally distributed with time has been described. (Ref. 6). The concentration is measured at truly random intervals and the values are plotted on logarithmic probability graph paper against the cumulative probability for each sample in a set of the total number of samples taken. The cumulative probability for individual samples, arranged in ascending order of magnitude, in groups of 5, 7, 10 and 12 random samples is shown in Table 9.2.

If the samples are drawn from a log-normal distribution of concentrations the plotted points will lie on a straight line and the parameters of the distribution can be determined as previously described. If they do not fall on a straight line a reason should be sought. For example, a change of conditions of operation may have occurred, or a new source of contamination may have arisen during the period when measurements were being made. If the time of the change can be identified it may be possible to analyse the data in two separate groups, before and after the change. It is not advisable to attempt to linearise the data by mathematical transformation.

TABLE 9.2   Cumulative probability for a number of random samples

Number of random samples	Designation of individual measurements											
	1	2	3	4	5	6	7	8	9	10	11	12
5	12.3	30.8	50.0	69.2	87.7							
7	8.9	22.4	36.3	50.0	63.7	77.6	91.2					
10	6.2	15.9	25.5	35.2	45.2	54.8	64.8	74.5	84.1	93.8		
12	5.2	13.1	21.5	29.5	37.8	46.0	54.4	62.2	70.5	78.5	86.9	94.9

RELATION BETWEEN VALUES AVERAGED OVER DIFFERENT TIMES

If the concentration of a contaminant varies with time, the average concentration will depend on the length of time over which the average is taken and the variation in concentration during that time.   The relation between the arithmetic means of a concentration when the average has been taken over various times has been studied for the special case when the concentration varies according to the log-normal probability distribution. (Ref. 4).

If the concentration of a contaminant is log-normally distributed with time and is averaged over a long time, T, (the criterion for a long time is considered later), then a mean value termed the Long Term Average (L.T.A.) is found.   If average concentrations are determined over a series of periods of short time, t, these short time averages will show a scatter about the L.T.A. (Fig. 9.4).

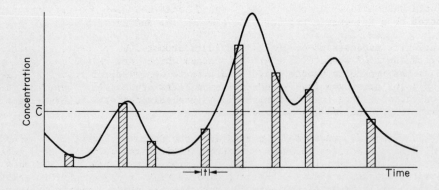

Fig. 9.4.   Concentration of a contaminant which varies log-normally with time.   The horizontal line at $\bar{c}$ is the mean, when an average value is taken over the whole time shown in the diagram.   The shaded areas are of short duration, t, and show the scatter in values found when the average is determined over these short times.

However, the actual values of the short time averages show considerable scatter both above and below the L.T.A. Using the fact that the instantaneous values of concentration are defined by the log-normal probability distribution, and by assigning a parameter obtained from experimental data, it has been shown (Ref. 4) how the scatter varies when the short time averaging interval is changed. In particular, it has been shown that 95% of all the values of the average measured over a short time, t, will fall below the value (R x L.T.A.) where R is a function of T/t. Values of R, calculated using an average value for the required parameter, are shown in Table 9.3.

TABLE 9.3   Values of R for Various Ratios of T/t

T/t	R
10	1.7
100	2.0
1000	2.25
10000	2.55

Thus, if $T = 100$ x t, then 95% of values averaged over time t will be less than twice the L.T.A. Alternatively, if 95% of values measured over time t are less than twice a specified limit, there is 95% probability that the L.T.A. will be less than the limit

The factor R is dependent on the ratio T/t and not on any particular unit of time. There is, however, a limit on the time which can be specified if the relation between the averages is to hold. The value of R is a measure of the variance of the short term averages. The "long time" chosen over which to specify the L.T.A. should be long enough for the variance on the measurement of L.T.A. to be approximately zero, that is, that on repeated measurements over the long time T the same value should be obtained for the L.T.A. Values obtained from measurement over times longer than T should still have the same L.T.A. and zero variance.

This work, which was carried out in Germany, was prompted by a desire to specify limits of exposure of workers to toxic substances which were values averaged over a year. It is impracticable to wait for a year to determine if a worker's exposure has exceeded the permitted level, and so measurements are made over shorter intervals and the relation is used to determine if it is probable that the exposure over one year will be within the permitted value.

The relation was also used in defining the European Economic Community Standard for exposure to vinyl chloride. By the time this standard was being prepared several countries within the Community had introduced national standards. These were based on different times of averaging, ranging from 8 hours to 1 year. The relation between the averages measured over the various times is used to permit various short time averages to be specified, all of which lead to the same long term average.

Ceiling Value

This relation between averages taken over various times has implications on the specification of ceiling values which are usually averaged over 15 minutes. If the concentration of the contaminant is accurately described by the log-normal distribution with the parameters used to obtain the value R tabulated in Table 9.3, and if the long time average over 1 year is specified, then 95% of values averaged over 15 minutes will fall below 2.7 x the annual average of concentration in the natural course of events, and 5% will exceed this value.   Thus the ceiling value specific- ation has no meaning under these conditions.   If it exceeds this calculated value it will always be complied with; if it is below this calculated value it cannot be complied with for 95% of the time.

Measurements Taken over very Short Times

The argument can be extended to values which are averaged over even shorter times than 15 minutes, but there becomes a limit beyond which the relation would not be expected to hold.   The difficulty in testing the relation at shorter times arises partly from limitations of methods of measurement, most of which have a finite time of sampling and measurement which is comparable with the averaging times concerned. While the concentration has a steady value this measurement time has no effect on the measured value.   However, if the concentration is changing rapidly the observed value depends on the measurement time of the analytical method.   For example, if a short peak of concentration is being measured the value of the maximum concentra- tion can be obtained only by a truly instantaneous measurement.   If the measurement is averaged over any time, the value obtained will depend on the shape of the peak. The shape of a peak of concentration depends on the source of the emission and the means of dispersal of the contaminant.   A large accidental emission would be expected to give a peak which shows rapid rise in concentration.   The decay of the peak would depend on whether the contaminant were dispersed only by diffusion or by wind or forced ventilation.   If for ease of calculation it is assumed that a peak shows a constant rate of increase to a maximum, followed by the same constant rate of decay, the peak will be triangular in form.   If the peak exists for a time, w, defined by the width at half maximum height, and is measured by a system which makes a measurement over a time, t, symmetrical about the maximum value, the result of measurement will be a function of the ratio t/w, and will have the percentages of the true value of the peak shown in the following table:

t/w	Measured value as % of true value of peak
2	50
1.5	62.5
1	75
0.5	87.5
0	100

This again illustrates the difficulty of comparing measurements made by systems which have different sampling and measurement times when the concentration being measured is changing rapidly.

The effect of sampling conditions on the measurement by colour indicator tubes of the average of a concentration which is changing during the period of sampling has been discussed. (Ref. 6).   It was shown that a bellows pump took 90% of its 100 ml sample in 18 seconds and the remaining 10% in a further 6 seconds.   A piston pump used in the same test took 90% of its 100 ml sample in 50 seconds and the remaining

10% in a further 70 seconds.   In consequence, peaks of concentration of short
duration can be seriously under-estimated, especially if the high values occur
during the period when the sampler is operating at low flow rate.   The particular
piston pump used in these tests is particularly susceptible to this error because
of its long period of low sampling rate.

REFERENCES

1. D.C. Stevens, The particle size and mean concentration of radioactive aerosols
    measured by personal and static air samplers, Ann. Occup. Hyg. 12, 33 (1969).

2. A.R. Jones & R.S. Brief, Evaluating benzene exposures, Amer. Inst. Hyg.
    Assoc. J. 32, 610 (1971).

3. N.A. Leidel & K.A. Busch, Statistical Methods for the Determination of
    Noncompliance with Occupational Health Standards.   National Institute for
    Occupational Safety & Health, Cincinnati, Ohio, U.S.A. (1975).

4. W. Coenen, Beschreibung das zeitlichen verhaltens von schadstoffkonzentrtionen
    durch einen stetigen Markowprozess, Staub - Reinhalt Luft, 36, 240 (1976).

5. R.I. Larsen, A new mathematical model of air pollutant concentration averaging
    time and frequency, J. Air Pollution Control Assoc. 19, 24 (1969).

6. K. Leichnitz, Air analysis at work places by means of short-term and long-term
    detector tubes, Draeger Review 42, 3, Draegerwerke AG, Luebeck, Germany, 1978.

Chapter 10

FUTURE POSSIBILITIES

Monitoring procedures have been developed over the years to provide the more frequent and more accurate measurements now called for to ensure the safety of workers and of residents who live around industrial plants. Monitoring places a burden on the plant management both in the provision of the specialised equipment and in the employment of technical staff. Developments have, therefore, followed three main lines; to improve the quality of measurement, to reduce operating costs, and to reduce equipment costs.

Present systems of measurement of most substances can produce information of the quality required by present systems of control. However, the methods of measurement available for certain substances are still not completely satisfactory in all circumstances, and stricter measures of control may be imposed on other substances, demanding more sensitive methods of measurement. Development of existing methods therefore continues and new and refined techniques can be expected. For example, a method recently proposed by NIOSH for the collection and determination of formaldehyde combines a chemical reaction with adsorption on charcoal. The charcoal is impregnated with a reagent with which the formaldehyde reacts when it is adsorbed on the charcoal. When the adsorbed substances are desorbed the formaldehyde appears as formic acid which is measured. However, simplification of measurement procedures is always desirable and it seems likely that simple direct-reading devices, such as those based on selective semi-conductors, will be developed and used more widely.

The main operating costs in monitoring are concerned with the employment of specialist staff in checking the performance of equipment and in sample handling and analysis. In particular, the systems of measurement of time-weighted average by sample collection on adsorbent traps of various types, which in its original form involved solvent desorption, was very demanding on analytical manpower. The development of thermal desorption methods has reduced this demand. The later development of colour indicator tubes for time-weighted average measurements has reduced the analytical requirement even further.

These systems, however, still require accurate metering pumps and these tend to be costly. They require frequent checking and maintenance because the accuracy of measurement depends on the quality of their performance. To eliminate these pumps the passive samplers based on permeation and diffusion have been developed. With the exception of certain devices which use specific chemical or electrochemical reactions to detect certain substances, most of these passive samplers still require

the services of the analytical laboratory for general applications.   The most
recent development is of a passive sampler which contains an adsorbent element which
is analysed by gas chromatography after thermal desorption.   This is probably the
minimum analytical effort demanded by any of the current design of samplers, and a
move to direct reading devices will almost certainly develop.   There has been
criticism of the present direct reading time-weighted average monitors, in that they
are open to misuse, and can be fraudulently exposed to show a credible over-exposure.
A direct reading device in which the accumulated reading is stored electrically
within the monitor, and can be read only when the monitor is attached to a reading
unit, would eliminate this possibility.   The personal monitor based on continuous
measurement by colour change on paper tape acts in this way, but more versatile
techniques will be developed.

Originally monitoring consisted of making measurements on grab samples.   Fixed
continuous monitors followed, but these monitored only the area immediately adjacent
to the sampling probe.   Elaborate sampling systems have been developed but they
remain point analysers in multiple form.   Some development has also taken place in
long path-length radiation analysers to give general surveillance of the atmosphere.
The application of these in factories, and the interpretation of the results, present
problems.   However, an analyser which would give general coverage of the working
area and identify points of potential hazard would be beneficial.

Personal monitoring was adopted to supplement the limited information provided by
working area monitoring.   With the exception of the permeation and diffusion
monitors, which are not yet fully accepted, these devices at present involve costly
equipment which is more or less inconvenient to wear.   It is not practicable,
therefore, to monitor all workers all of the time, as is done in situations of
radiation hazard.   Further, the monitor measures the concentration to which the
worker is exposed, which is the parameter expressed by the hygiene standards.
Neither the monitor nor the limit take any account of the different types of work
on which operators may be engaged, and which may affect their susceptibility to the
contaminant.   The amount of toxic substance absorbed from inhaled air is given
approximately by the equation:

$$T = \int V(C_i - C_e)dt$$

where T =   amount of toxic substance absorbed

V =   rate of inhalation of air

C_i =   concentration of toxic substance in
                inhaled air

C_e =   concentration of toxic substance in
                exhaled air.

V can vary greatly and is a function of work load, and in consequence the amount of
toxic substance absorbed is dependent on factors other than atmospheric concentration.
Systems of biological monitoring are being developed in which the amount of toxic
substance absorbed by a worker will be measured by the composition of exhaled air,
urine or body fluids.   The exact procedure will vary from one substance to another
depending on whether the substance is retained in the body in its original form or
metabolised, and at what rate.   Systems based on the analysis of exhaled air are

already in use for the measurement of consumption of alcohol and of exposure to
carbon monoxide and benzene vapour, and will be developed for general application
to other substances.

The present methods of sample collection and measurement are, in general, adequate
for current needs.   However, as allowable exposures are reduced, and as more
strict control measures are introduced to reduce the risks from toxic chemicals,
the levels of concentration which have to be monitored will become much lower.
Then it may be necessary to question the efficiency of certain collection methods
and to develop new methods of sample collection which will perform adequately at
these very low values of concentration.   Analysts have shown great ingenuity in
applying a wide range of techniques to the detection and measurement of atmospheric
contaminants and will continue to apply new techniques of measurement to this
problem as new methods are developed by analytical science.

It may be asked how far the pursuit of lower levels of exposure, and the consequent
demand for more sensitive methods of measurement, may be expected to go.   Risks
from chemical toxicity must be brought down to a level at which they are acceptable
to society.   This may be a level at which they present a risk which is not greater
than that presented by other equivalent hazards.   However, if it can be shown that
a given concentration of a substance may, if breathed, interfere with human
metabolism in such a way as to cause an adverse effect on health within the human
life-span, it is likely that there will be demands for exposure to be limited to
concentrations below that level.   This, of course, implies that the contaminant at
that concentration can undergo some reaction with a body chemical, and the study of
these biological reactions may well lead to new measurement techniques of high
sensitivity.   Such reaction may also show very high specificity, not for individual
compounds, but for types of compounds which cause particular effects.   It may then
be possible to develop tests which are specific for compounds which are toxic in
certain ways, or which are carcinogenic, and so relieve the analyst of having to
monitor the presence of individual compounds.

SELECTED RECENT RELEVANT
PUBLICATIONS

CHAPTER 1

(1) J. Manos, Industrial survey-chemicals, *Health & Safety at Work, 1 (11),* 28 (1979).

Discussion of procedures used for setting hygiene standards for chemicals.

CHAPTER 2

(2) D. J. Ball and M. J. R. Schwer, New developments in air pollution monitoring techniques, *Clean Air, 8, No. 31,* 25 (1978).

Report on a conference held in London in October 1978 on lasers and other advanced optical techniques, e.g. Lidar, correlation spectroscopy, long-path infra-red, for pollution monitoring.

(3) G. F. Kirkbright, Spectroscopy by light and sound, *Bulletin of Nation Research Development Corporation, 49,* 20 (1979).

Description of technique of opto-acoustic spectroscopy.

(4) G. Schunck, Non dispersive infra-red gas analysers for industrial processes and protection of the environment, *Measurement and Control, 11,* 245 (1978).

Description of a medium path length analyser.

CHAPTER 3

See Refs. (7) and (14).

CHAPTER 4

(5) Sieger Autospot, *Internat. Environm. & Safety,* 4 (Aug. 1979).

Description of a new semi-automatic device for measurement of instantaneous atmospheric concentrations by colour change of impregnated paper tape.

(6) K. Leichnitz, Some information on the long-term measuring system for gases and vapours, *Draeger Review*, *43*, 6, Draegerwerk AG, Luebeck, Germany (1979).

   Properties of long-term colour indicator tubes for measurement of time weighted average concentrations.

(7) K. Leichnitz, How reliable are detector tubes?, *Draeger Review*, *43*, 21, Draegerwerk AG, Luebeck, Germany (1979).

   Description of inter-laboratory tests of indicator tubes, statistical analysis of the results, and evaluation of the properties of production batches.

   CHAPTER 6

(8) D. W. Gosselink, D. L. Braun, H. E. Mullins and S. T. Rodriguez, A new personal vapour monitor with *in-situ* sample elution, *Annual Conference of Amer. Ind. Hyg. Assoc.*, Chicago, May 1979.

   Full description of the personal sampler mentioned in Chapter 6, Ref. 8.

(9) Mercury Vapour Dosimeter Badge, Bastock Marketing, Aynho, Banbury, Oxon, U.K.

   A passive sampler which contains a treated charcoal which readily absorbs mercury vapour. The mercury is subsequently desorbed and measured by flameless atomic absorption.

(10) Personal monitoring techniques for gases and vapours, *Internat. Environm. & Safety*, 43 (Apr. 1979).

   Description of carbon collection, thermal description system for personal monitoring.

(11) J. Walton, Personal sampler for TWA measurements of toxic gases and vapours, *Internat. Environm. & Safety*, 29 (Aug. 1979).

   Review of techniques of personal monitoring.

(12) D. Coker, Trends in personal monitoring techniques, *Protection*, 5 (July 1979).

   Recent developments in personal sampling.

(13) GN Concentrator - An automatic two stage desorption and injection device, GN Instrumentation Consultancy Ltd., Wimbledon, London, U.K.

   An automatic thermal desorber with a cryogenic trapping stage to concentrate the sample.

   CHAPTER 7

(14) W. Thain, Performance requirements of pumps for long-term sampling, *Internat. Environm. & Safety*, 33 (Aug. 1979).

   Description of the errors which can arise in time weighted average measurements due to errors in the pumping system.

CHAPTER 8

(15) J. H. Scawin, The stability of calibration gas mixtures, *Internat. Environm. & Safety*, 25 (Apr. 1979).

Factors which affect the stability of calibration gases when stored in metal cylinders.

(16) J. H. Clements, Blending calibration gases, *Internat. Environm. & Safety*, 23 (Jun. 1979).

Apparatus for blending gases by a dynamic system using calibrated flow regulators.

(17) The UP-37 Permeator - An ultra-portable, battery operated unit for on-site calibration, CEA Instruments, Inc., Westwood, New Jersey, U.S.A.

A portable device which contains a permeation tube and a constant flow of compressed gas which are maintained at constant temperature. The primary standard which is generated is suitable for calibrating any type of analyser.

GENERAL

(18) D. C. M. Squirrell and W. Thain, *Environmental Carcinogens, Selected Methods of Analysis, Vol. 2, Methods of Measurement of Vinyl Chloride in Poly (Vinyl Chloride), Air, Water and Foodstuffs*, I.A.R.C. Scientific Publication No. 22. Internat. Agency for Research on Cancer, Lyon (1978).

Standard methods for the determination of vinyl chloride. Also contains a critical review of methods proposed and used for measurement of vinyl chloride, many of which are suitable for measurement of other chlorinated compounds.

(19) *NIOSH Manual of Analytical Methods, Vol. 5*, U.S. National Institute for Occupational Safety and Hygiene, Cincinnati, U.S.A. (1979).

An addition to the NIOSH manual containing a number of newly developed methods of analysis.

(20) W. Thain, Monitoring toxic gases in the atmosphere, *Occupational Safety and Health*, 9 (10), 10 (1979).

A review of monitoring methods.

SUBJECT INDEX